A gift for

_____

from

_____

I Used to
Know That

I Used to Know That

GENERAL SCIENCE
stuff you forgot from school

MARIANNE TAYLOR
Foreword by Caroline Taggart

MICHAEL O'MARA BOOKS LIMITED

This book is for my family.

First published in Great Britain in 2010 by
Michael O'Mara Books Limited
9 Lion Yard
Tremadoc Road
London SW4 7NQ

A CIP catalogue record for this book is available from the British Library.

Papers used by Michael O'Mara Books Limited are natural,
recyclable products made from wood grown in sustainable forests.
The manufacturing processes conform to the environmental regulations
of the country of origin.

ISBN: 978-1-84317-473-8

1 3 5 7 9 10 8 6 4 2

www.mombooks.com

Typeset and designed by Design 23

Front cover lettering by Toby Buchan

Printed and bound in Great Britain by Clays Ltd, St Ives plc

CONTENTS

Foreword by Caroline Taggart   9
Introduction   13

∿

PHYSICS

Energy and electricity
    Generating electricity   18
    Heat transfer and efficiency   23
    Using electricity   27

Forces
    The four fundamental forces   32
    Planets, stars and galaxies   35
    The origins of the universe   40
    Laws of physics   42

Waves, radiation and space
    Waves   45
    The electromagnetic spectrum   48
    Radioactive substances   53

∿

CHEMISTRY

The periodic table
 How the table works   58
 Atomic structure   63
 Chemical bonds   67
 Chemical reactions   73
 Collision theory and rates of reaction   78

Fuels, air, pollution
 Chemicals in the air   83
 Measuring pollutants   87
 Useful chemicals from crude oil   88
 Making life cycle assessments   90

Metals
 The Earth's structure   93
 Metals and alloys   96
 Construction materials   98

Organic chemistry
 Natural polymers and their roles in nature   102
 Nutrition   107
 Harmful chemicals   110

BIOLOGY

Human (and other) bodies
    Circulation   114
    Skeletal structure   119
    Muscles and skin   122
    Nervous system   126
    Digestive system   130
    Reproductive system   133
    Respiratory system   137
    Sensory systems   140

Cell biology
    Structure of a cell   146
    Photosynthesis   150
    Hormones   154

Evolution and environment ecology
    The origins of life   157
    The evolution of the eukaryotic cell   160
    Mutation and natural selection   163
    Population   169
    Predation   171
    Extinction   175

Genetics
    Chromosomes   178

Inheritance   180
Reproduction and cloning   186

Index   188
Further reading   192

FOREWORD

When the original version of *I Used to Know That* was published two years ago, I spent a very jolly couple of days in a small BBC studio in central London. With headphones over my ears and a microphone in front of me, I talked to people on radio stations all over the country about the book: why I had written it, what they liked about it and what brought back hideous memories.

To my surprise, the hideous memories were what excited people most. Top of the list – and this bit *wasn't* a surprise – was maths. One listener said that just looking at the letters a + b = c on the page had brought him out in a cold sweat, even though he no longer had any idea why. Another radio station carried out a series of interviews in the street asking people, among other things, if they knew who Pythagoras was. 'Oh yes,' said one man, 'he's to do with triangles and angles and all that malarkey.'

I thought that was wonderful: 'all that malarkey' summed up perfectly the way many of my generation were taught. We had to learn it (whatever 'it' was); we were never really told why; and, once exams were over, unless we went on to be engineers or historians or something similarly specific, we never thought about it again. But it lingered somewhere at the back of our minds, which may be why *I Used to Know That* touched a chord.

However, covering five major subjects and including a catch-all chapter called General Studies meant that a single small volume couldn't hope to deal with anything in much depth. This is where the individual titles in this series come in: if *I Used to Know That* reminded us of things that we learnt once, these books will expand on them, explain why they were important and even, in the case of *Science*, update us on things that have been invented and discovered since we went to school. If you enjoy this one, look out for *I Used to Know That: English, Maths, History* and *Geography* as well.

Science was, I have to confess, far and away my worst subject. I remember only the oddest things: being dazzled by the beautiful blue of a copper sulphate solution and fooling around with a heap of iron filings and a magnet. I knew about Vitamin C (oranges, for avoiding scurvy) and Vitamin A (carrots, good for the eyesight) long before anyone had invented the concept of 'your five a day'. I dare say I also knew that the knee bone was connected to the thigh bone, but that was about as far as I got.

That's why this book is so fascinating. Marianne Taylor has not only explained lots of topics that many of us may only half remember or never really have understood; she has also expanded on things that are all around us – things that we take for granted but that are a fundamental part of our lives on Earth (or indeed on other planets, should we ever find ourselves there). The air that we breathe;

the electromagnetic waves that power microwave ovens and x-ray machines; the way our genetic make-up turns us into the individuals that we are; those unimaginably small atoms that can be split to unleash devastating power – these are all elements (no pun intended) in that huge subject called 'general science'.

If you enjoyed science at school, this book will remind you why. If you didn't, it will show you what you have been missing and help you to catch up. It will clarify some of the scientific arguments that appear in newspapers today and make you realize that an enthusiasm for science doesn't (necessarily) make you a nerd.

And if you have some fun along the way, so much the better.

CAROLINE TAGGART
LONDON, 2010

11

INTRODUCTION

Most of us know there's fun to be had from science, as well as an immense sense of achievement when you do finally grasp a tricky concept, but our experiences from school days may put us off trying, and we switch off our brains the moment things get 'too technical' – school's finished and there's no need to struggle with this stuff any more, right?

Well, when you leave school, you can indeed leave behind a lot of the stuff you learned. You can go day after day, week after week without needing to remember the names of the kings and queens of England or the world's tallest mountains and longest rivers. You may never need to speak French or German in your life again, and if your command of written English goes downhill somewhat, it probably will go unnoticed by most (more's the pity). Science is different, though. Science is everywhere, it muscles in on your day-to-day life whether you want it to or not. From deciding which vitamins you should take to working out how best to heat your home, it's all about scientific principles, and you have everything to gain and nothing to lose from familiarizing – or re-familiarizing – yourself with a bit of science.

This book covers more or less the same stuff you'll find in an average GCSE combined syllabus. However, we're not trying to get you through an exam but to get you excited

THE SCIENCE TRINITY

The traditional three scientific subjects taught at school are physics, chemistry and biology. When I did my GCSEs they were taught as separate subjects, and you chose one, two or all three. Most GCSE students today study 'combined science' which counts as a double GCSE, which might seem like a step in the wrong direction. However, it makes good sense on one level, in that separating out the sciences is difficult (and actually not all that sensible) because there is much overlap and understanding aspects of one relies upon having a grasp of aspects of the others.

Physics is the study of fundamental forces and particles. Because it deals with things that we mostly can't see directly, there's a lot of theory and a lot of mathematics and therefore it's the one science that people generally find the most daunting. Physics is sometimes considered the study of the very big and the very small, from universes and stars to subatomic particles.

Once you have assembled enough subatomic particles to make an actual atom, you enter the realm of chemistry. This science deals with the properties and behaviour of atoms and molecules of the many and varied elements and compounds, in various different settings from the school lab to the blast furnace, from the atmosphere of

Earth to the insides of the cells in our bodies.

And so chemistry segues effortlessly (well, more or less) into biology, the study of living things. From the molecular machines that drive our internal processes, biological study works up to the structure and organization of the cell. Then on to the ways cells of different types are combined to form the body's various different systems, how those systems work, and finally from the individual organism to the ways populations and whole ecosystems work.

about science, so here and there I may have said a bit more about some things and a bit less about others. Some of the concepts we're going to explore are difficult, but every care has been taken to explain stuff in straightforward language and to steer clear of unnecessary terminology, mind-bending mathematics and overly esoteric subject matter. Wherever possible, diagrams and real-world examples are added to help make things more understandable.

You may find after you've read the book that you're hankering after a bit more detail about certain topics – if so, please check out the Further Reading section at the end for ideas on where to go for more in-depth study. If the sciences didn't work the way they do we wouldn't be here, and being able to understand how science works is one of the best things about being human.

SCIENCE AND PSEUDOSCIENCE

Because it applies to everything in this book, this seems like the right place to explain what the scientific method is – and isn't. For science to work, it needs to be carried out with an honest and consistent method. The basic process is that a hypothesis (a possible explanation) for a phenomenon is proposed. Then we come up with predictions that would fit that hypothesis, and test the predictions to see if the hypothesis can be proven false.

Pseudoscience covers things that look science-like but don't fit the scientific method. For example, we can't scientifically test for the existence of a god, because the statement 'God exists' can't be falsified – we can put down any lack of evidence to supernatural explanations – God is omnipotent and can therefore choose to go undetected regardless of our searching methods. Pseudoscience also describes things which are claimed to be true but for which convincing evidence has not been found.

PHYSICS

ENERGY AND ELECTRICITY

GENERATING ELECTRICITY

Electricity has to be up there as mankind's favourite energy source. It's easy to work with and we can use it to power all kinds of machines and processes. However, naturally occurring electricity is rare and unpredictable – harnessing lightning strikes isn't really a practical option, for example. So we generate our electricity by converting other, more easily obtainable, energy sources.

Natural Energy Sources

Our world is replete with many different potential sources of energy, which we can convert into our preferred form – electricity. Some of them are sustainable – we can keep on using them indefinitely; others aren't – we'll use them up eventually and they won't get replaced.

Fossil fuels are coal, natural gas and oil. They are called

fossil fuels because they are the compressed remains of ancient forests (coal) or tiny sea animals (gas and oil), squashed up and broken down into carbon-based compounds which burn readily, releasing heat. They are not sustainable because the conditions that produced them no longer exist. Also, burning them releases extra carbon dioxide and other unwelcome gases into Earth's atmosphere, contributing to global warming. Despite that, about 75% of the electricity used in the UK comes from power stations that burn fossil fuels, though alternative sources are becoming more and more widely used.

Nuclear power is generated by breaking up atoms of plutonium and uranium. When nuclei of atoms from these two elements are battered with particles called neutrons, they release heat energy. Inside the nuclear reactor of a nuclear power station, this process (nuclear fission) is initiated and controlled, and the energy harnessed. The process doesn't release toxic gases, but the waste it generates is dangerous. Nuclear fuel is non-sustainable, and an accident at a nuclear power station could have very far-reaching and disastrous consequences.

Wind power is a sustainable source of energy. Large wind turbines, like giant futuristic windmills, are sited in exposed places with strong prevailing winds. The wind turns the turbine's blades, and this kinetic (moving) energy is captured. No pollutants are produced, but wind farms can be unsightly, noisy and dangerous to birds.

E=mc²

Bet you weren't expecting to see this quite so soon. This most famous formula, from Einstein's Theory of Special Relativity and first proposed in 1905, is not as scary as you might think, however. It just means that matter (any kind of physical substance – like rock, a star, a bar of chocolate or your cat) and energy are different versions of the same thing. This comes into play in nuclear fission: the products of the fission reaction together weigh slightly less than the original fission material because the 'missing mass' has been converted into energy.

> E = energy
> m = mass (measured in grams and kilograms)
> c = the speed of light (which is a constant,
> i.e. it doesn't change no matter what)

The formula shows us how much energy is stored in a quantity of matter. The main points to understand from this are that a) matter is a kind of energy, and b) energy doesn't go anywhere. You can't add new energy to the universe and you can't take any away, but you can change it from one form to another. What this has to do with generating electricity is that we can convert any form of energy into any other form of energy without 'losing' any of it. Some of it might escape from our control but if our energy-converting machinery is well-designed we can keep this to a minimum.

Water power is another sustainable energy source. Tidal barrages across estuaries capture kinetic energy as the tidal changes move water in and out of the river mouth. Hydroelectric power stations use dams built across rivers and capture the water's energy as it flows down pipes built into the dam. Neither produce pollution, but they damage wildlife habitats.

Geothermal power stations capture the natural heat energy from volcanic areas. It is sustainable and does not produce pollutants, but is only available in a few places worldwide.

Other energy sources include solar power, wood, biofuels (oils derived from crops which can be burned like fossil fuels) and hydrogen fuel cells. The energy sources used by a country vary quite a bit. For example:

UK
74% fossil fuels, 19% nuclear power, 1% hydroelectric power, the rest from other renewable sources or imported

USA
83.4% fossil fuels, 8.5% nuclear power, the rest from renewable sources

Iceland
74% geothermal power, 25% hydroelectric power, 1% fossil fuels

Worldwide
we use 87% fossil fuels, 6% nuclear power, 6% hydroelectric power and 1% A N Other. Expect these figures to have changed quite a bit in fifty years' time.

Whatever your source of energy, the process of converting it to electricity requires that the energy first be made into kinetic energy, and then that this kinetic energy is used to drive a generator to produce the electricity. In the case of wind and water power, the harnessed energy is kinetic already. With the others, heat energy must be converted to kinetic energy – usually by using the heat to turn water into steam, which then powers a turbine of some kind. The generator has an electric charge in its workings already, but it's the input of kinetic energy that causes the electric charge to flow.

HEAT TRANSFER AND EFFICIENCY

As I write, it's winter time and we've had some problems with the boiler. Moving heat energy around in the most efficient way possible is a major human preoccupation, and if we'd never learned how to do it we wouldn't have been able to colonize such a large proportion of the available land surface on Earth.

Heat has three ways of getting around: conduction, convection and radiation.

Conduction means travelling directly through matter, whether solid, liquid or gas. If you hold a spoon over a candle flame, pretty soon you're going to want to drop the spoon because it's too hot to hold – that's conduction. Some materials conduct heat better than others – the structure of metals makes them good conductors so we make radiators out of them.

Convection is the movement of heat through liquids and gases. When you add extra hot water to your bath, the heated water expands (or to be precise the spaces between the water molecules get bigger as the molecules start moving about more), becoming less dense and thus taking up more space. It rises above the more dense cooler water, which sinks. The rising of hotter, less dense liquids or gases and falling of cooler and more dense liquids or gases produces convection currents. You might see birds of prey taking advantage of the rising hot air currents – thermals – that

THERMODYNAMICS

Time to lay down the law. Or laws in this case. The movement of heat is governed by four physical laws, which are as follows (converted into sensible language).

THE ZEROTH LAW*
'If two heat systems are both in equilibrium with a third, they are also in equilibrium with each other.'

THE FIRST LAW
'In a closed system, the amount of heat remains constant.'

THE SECOND LAW
'In a closed system, heat will always spread out as evenly as it can.' (Another way to put this is that its entropy or state of disorderedness will always increase.)

THE THIRD LAW
'At the temperature absolute zero, the entropy of a system is zero.'

---

*It's 'zeroth' because it was added as an afterthought, being almost too obvious to be worth writing down.

The Second Law is the one of interest here, because it's what we are working against when we try to keep our homes warmer (or cooler) than the world around us.

form by hillsides to gain height effortlessly. So convection describes the tendency of heat to go up, and is the reason we don't put our room heaters on the ceiling.

Radiation is a bit different. Conduction and convection describe the transfer of heat through the particles that make up matter, whether solid, liquid or gas. Radiation comes in waves, not particles. It's a part of the electromagnetic spectrum that describes all kinds of waves, including microwaves, visible light waves, x-rays and radio waves. We'll be looking at waves properly later on – for now all you need to know is that infrared radiation = heat. Every kind of matter absorbs and gives off (reflects) infrared radiation. Things with more surface area for their volume (long and narrow) absorb and reflect more than things with less (short and wide). Objects that are white and/or shiny absorb and reflect more than those that are dark and/or matt. Why are radiators painted white, then? Because, despite their name, they work mostly by convection and conduction.

All of these three modes of transport can be used to

heat up the rooms in our homes, but need to be blocked as much as possible to keep the heat in where we want it, rather than letting it escape into the wider world. Those who live in the tropics have the same problem in reverse – outside heat needs to be kept out if it's too excessive. For this reason we need to use insulation – materials that discourage conduction, convection and radiation through our walls, windows and roofs.

Because of convection, buildings lose most heat through the roof, and all the materials that the building's

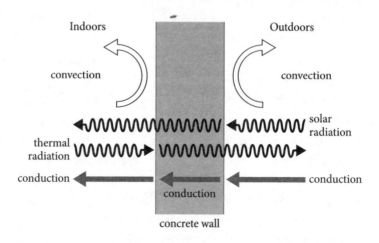

outer shell is made from can conduct heat out of (or into) the building. Therefore we need to use materials that are good insulators (bad conductors, in other words) when we make our buildings. The fluffy padding in loft and cavity wall insulation is a good example – it is a poor conductor and it also traps pockets of air. Still air is a poor conductor too, and by preventing air circulation you also reduce heat loss by convection. Double glazing is another way of trapping air, and blocking draughts has a similar effect.

USING ELECTRICITY

Electricity is immensely useful but also potentially dangerous. For us to be able to use it safely in our homes it needs to be at an appropriate voltage. However, large-scale transportation of electricity requires a much higher voltage, which is itself lower than that of electricity fresh from the power station generator.

MEASUREMENTS OF ELECTRICITY

Electric current is measured in amperes (amps for short) – an expression of the amount of current flowing through a conductor over a specified timeframe. So what about volts? A volt is a measurement of potential difference – the difference in electric charge between two points. The negatively charged particles – electrons – that make up an electric current flow from areas of high negative charge to areas of low negative charge (in other words areas of high positive charge). Voltage can be calculated by dividing power (in watts) by amps. The formula looks like this:

volts = watts/amps

or this:
watts = amps x volts

or indeed this:
amps = watts/volts

To help get your head around this, imagine water flowing through a hose attached to a tap. The rate of flow (amps) can be made more powerful (watts) by turning the taps on more fully to put more pressure on the water (volts).

CHANGING VOLTAGES

We've already seen how generators in power stations produce electric current. To move the electricity around the country, we need to give it a low current to minimize energy loss – high currents running through wires lose energy in the form of heat. A low current means a

SOME VOLTAGE FACTS

Our power stations churn out electricity at a voltage
of 25,000 volts. The step-up transformers increase
this to between 132,000 and 400,000 volts, for
distribution through the National Grid. Step-down
transformers reduce the voltage to a considerably
less hazardous 230 volts for use in our homes.

high voltage. We use transformers* to produce the high
voltages necessary, and because such voltages are very
dangerous, the design and siting of pylons and power
lines needs to be planned very carefully.

When the electricity goes into our homes, the voltage
has to be dropped to a safer level, so transformers come into
play again. These are called step-down transformers – the
ones that increase voltage are step-up transformers. Both
kinds use magnetic fields to alter voltage.

---

*Not to be confused with the robot/car toys, which are less useful though
arguably more fun, these transformers transfer electricity from one circuit
to another through coils of wire, which are wound around an iron core,
stepping the current up or down as required.

Resistance

Resistance isn't futile when you're talking about electricity, far from it. Lots of things can impede the progress of an electric current, and sometimes we use resistors on purpose because they change voltages. The equation looks like this:

voltage = current x resistance (measured in ohms)

In other words, when a current meets resistance, it needs more voltage to overcome that resistance.

Using Electricity In Your Home

We know that different appliances use different amounts of power and affect the size of our quarterly bills. The energy an appliance uses is measured in units called kilowatt hours (kWh) – it's calculated by this formula:

energy used (kWh) = power (kilowatts) x time (hours)

If you also know the cost of your electricity per unit you can work out how much it costs to run each appliance for an hour by multiplying the number of kilowatt hours by the cost per unit. Of course, appliances use different amounts of energy in different 'modes' – a laptop uses more power when in use than when on standby, for example, so for a really accurate calculation you'd need to factor that in too.

FORCES

THE FOUR FUNDAMENTAL FORCES

We live in a universe which is subject to a whole host of physical laws, without which neither we nor the world we live in would exist as we know it. The same is true of the four basic forces, which act upon matter and determine its behaviour. They are the electromagnetic force, the strong nuclear force (or strong force), the weak nuclear force (or weak force) and the force of gravity.

ELECTROMAGNETIC FORCE

Let's kick off with this one, as we've already had a bit of a look at electricity. This force is all about the attraction between dissimilar charges, and the repulsion between negative charges. Electricity is the flow of electrons – negatively charged particles. Electrons are also a constituent of atoms, as are positively charged protons. The electromagnetic force is what holds atoms together, and also molecules (two or more atoms bonded together – more about this in the Chemistry section). Moving charged particles create magnetic fields – so magnetism is related to electricity and together they form this force. It is the second strongest of the forces and operates over infinite distances.

The Nuclear Forces

The strong and weak forces both act on the particles inside the nucleus of an atom – the positively charged protons and uncharged neutrons. The strong force holds the protons and neutrons together, and is the strongest of the four forces. However, its range is small – about the size of an average atom's nucleus. The weak force is the second weakest of the four, and operates over a very small range – less than the size of a proton. It acts upon particles even smaller than protons and neutrons, and is responsible for nuclear decay – the breaking down of very large nuclei into smaller ones – also, without it the nuclear fusion that makes stars burn would not happen. That's a pretty good example of the fundamentality of these forces.

Gravitational Force

This is the weakest of the four forces. That may come as a surprise, given that gravity is the one that we are probably most intimately aquainted with, and we know just how firmly it insists that we don't float off into space, but there it is. It does make up for this feebleness by operating over infinite distances. Gravity is the force that attracts particles or larger bits of matter towards each other.

WEIGHT AND MASS

When we talk about something's weight, we're quite likely to also be talking about kilograms or grams, stones or ounces, or tons. These measurements, though, are units of mass rather than weight. Weight is a bit more complex, describing the gravitational forces an object is experiencing due to its mass. The equation looks like this:

weight = mass x g

The '*g*' is an expression of the local gravitational pull, which you'd determine by measuring how quickly a free-falling object accelerates. On Earth, *g* is about 9.8 metres a second. It is written as an italic so you don't get mixed up between it and the non-italicised 'g' which is short for 'gram'.

As we mostly don't take objects off planet Earth

in order to weigh them, the terms 'weight' and 'mass' are used interchangeably. However, when we escape the gravitational pull of this planet, the weight of our chosen object changes. In space, far from all other matter, it will essentially be weightless (but not quite – remember that gravity's reach is infinite so there's nowhere in the universe where there's no gravity at all) as anything multiplied by zero is zero. Take the object to the Moon, and it has some weight, but less than it does on Earth because the Moon is less massive. However, the object's mass doesn't change at all, ever, whether you take it to Basingstoke or Betelgeuse.

PLANETS, STARS AND GALAXIES

Our universe holds a whole lot of space, and concentrated bits of matter in the form of the various 'heavenly bodies'. Planets orbit stars to form solar systems, and stars are clustered into galaxies which usually show an ordered structure and movement. Our galaxy is the Milky Way, and our Sun is one of at least twenty billion stars within it. There are some hundred billion other galaxies out there in the observable universe. These are some seriously big numbers we're throwing around. One more biggie before we look at things in more depth – the observable universe is at least forty-six billion light years in any direction. And there is plenty more universe in the unobservable category too.

Planets

Back to Earth (or thereabouts) for a moment. Our solar system contains two kinds of planets – the solid rocky ones (Mercury, Venus, Earth and Mars) and the gas giants (Jupiter, Saturn, Uranus and Neptune). Rocky planets are made of … rock. The gas giants are much bigger than the rocky planets and are composed of gases. The outer two planets, Uranus and Neptune, are in a subgroup of gas giants called ice giants – they are less massive than the true gas giants and contain more solid material.

MOONS AND OTHER LITTLE BITS

In our solar system, only Mercury and Venus have
no moons. We have just the one, Mars has two small
ones (Phobos and Deimos), but Jupiter, the biggest
planet, has a whopping sixty-three of them (though
only four are of a comparable size to our moon).
Saturn has sixty-one, of which sixty are dwarfed by
the huge Titan, which is nearly twice the size of our
moon. Then things fall away somewhat – Uranus has
twenty-seven moons and Neptune just thirteen.

Our solar system also contains quite a lot of
extra matter in the form of asteroids or planetoids
(which is often used to refer to larger asteroids).
These are smallish (though sufficiently large to
do very significant damage to any planet that gets
in their way) lumps of rock that orbit the sun,
mostly in a belt between Mars and Jupiter. The
rocky Pluto, further out than Neptune, was once
designated a ninth planet, but is now considered a
large planetoid. Smaller rocks are called meteroids.
Comets are collections of icy and rocky debris,
usually with a 'tail' of gas and dust streaming out
behind them.

Stars

Stars are made of plasma. Not the same stuff that's in your blood, but a gas that consists partly of charged particles – both positive and negative. In stars, plasma is held together as a big ball by the force of gravity. In the case of our sun and many other stars, the centre of the ball is undergoing constant nuclear fusion reactions, whereby hydrogen gas is turned into helium. The energy this generates is radiated out from the star in the form of light and heat.

Every star has a lifespan that begins with its formation from a cloud of mostly hydrogen, and then passes through the fusion stage. What happens next depends on the size of the star. Medium-sized 'yellow dwarf' stars like our Sun will, on finishing up their hydrogen supplies, turn into 'red giants', their atmospheres expanding massively as their cores collapse under their own gravity. Our sun will do this in about five billion years' time – we will have needed to find a new solar system to live in at least two billion years before this, if we manage to dodge all the other potential catastrophes out there in the meantime. The red giant stage proceeds to the 'white dwarf' end stage, when the star sheds its atmosphere and becomes a very dense and much smaller and dimmer body.

If the star is too small for the nuclear fusion

process to begin, it's a 'brown dwarf' or a failed star. Stars intermediate in size between brown dwarfs and yellow dwarfs are red dwarfs, and they live billions of years longer than yellow dwarfs before becoming white dwarfs without passing through a red giant stage.

Very massive stars tend to live fast and die young, in a spectacular gravitational collapse producing a supernova – an extremely powerful explosion. A white dwarf star can also go supernova if it draws in enough extra matter. What's left may then become a very hot and very dense but stable neutron star – the biggest become black holes, which are anything but stable. Black holes draw in everything in their vicinity, like a whirlpool, and not even light can escape them. There is thought to be a supermassive black hole at the centre of our galaxy, and indeed at the centres of galaxies generally.

Gravity is what holds together the individual objects in the galaxy, and also what keeps moons orbiting planets, planets orbiting stars and galaxies spinning around their centres. Objects stay in orbit around bigger objects if they are moving fast enough that their forward motion is in balance with the gravitational pull of the bigger object – in the vacuum of space, there's no drag from an atmosphere to slow the object down so its speed stays constant.

THE ORIGINS OF THE UNIVERSE

Perhaps the biggest question science has to answer is where the universe came from in the first place. It's easy to see how early civilizations thought it was all about us. Before we knew the nature of the sun and moon, and the thousands of stars across the night sky, why would we not think our own world was the biggest, most important object in existence and the centre of everything? Now we know better, of course, and have come to terms with the fact that we are specks living on a speck orbiting a small star in the corner of a colossal galaxy of billions of such stars, among billions of other galaxies. And in between all that stuff are quite mind-bending quantities of empty space.

EXPANDING UNIVERSE

Scientists looking at distant stars have noticed something called the redshift. The light from distant stars is shifting to the 'slower' red end of the visible colour spectrum rather than the blue end. This means that the stars are moving away from us, and each other. It's the same phenomenon that makes police sirens rise in pitch as they approach us, and then lower in pitch when they (hopefully) go past and off into the distance – the Doppler effect.

This is the main evidence for the idea that the universe is currently expanding. That doesn't mean that the actual stars and other bits of matter are getting bigger, but that the space between galaxies is increasing. (It's also important to understand that the universe isn't expanding into 'empty space', the universe *is* (mostly) empty space.) That this is happening is strongly suggestive of the idea that the universe was once very much smaller, and that it had a defined beginning.

THE BIG BANG

Once there was nothing at all, then there was a big explosion and there was everything. Well, not quite. But that's quite a widespread (mis)understanding of what's meant by the Big Bang. I'm going to try to clarify things somewhat. Here's what the current theory says – about 13.7 billion years ago, the universe was in an extremely hot and dense state and had a very small volume – the same as an atom – but very rapidly expanding and cooling. The idea that an atom-sized object could hold all the matter that's in the universe today is, obviously, mind-bending.

What happened *before* the Big Bang, we just don't know. But we do know that what we define as the Big Bang wasn't an explosion, rather a massive and extremely rapid expansion. As things cooled down, the matter began to coalesce, under the influence of the fundamental forces, with free protons and neutrons coming together to form the nuclei of helium atoms, and hydrogen atoms forming from protons and electrons. The coalescence of gas clouds formed star systems, with planets formed from the heavier elements that are blown off from the outer layers of stars as they burn. *Et voilà*, one universe.

WHAT NEXT?

The universe is expanding now, but will it expand forever? Scientists are divided on this. Some say yes, some say no, it will slow down and reach a steady state, and some say that it will eventually begin to contract instead, and all matter and space will return back to its original hot, dense and tiny state in a Big Crunch.

LAWS OF PHYSICS

This seems like the right time to summarize some of the very numerous laws of physics that you will have encountered in school. Not all of the laws are described here – others are shown in more detail in the relevant sections.

LAWS OF CONSERVATION OF ENERGY AND MASS

Newton's three laws of motion talk about the relationship between an object and the energy forces it experiences.

The object remains at rest, or moves in a straight line, until a force acts upon it.

The acceleration of the object is proportional to the force causing it, e.g.:

force = mass (of the object) x acceleration.

When object A applies a force to object B, this action produces an equal and opposite reaction in object A.

LAWS GOVERNING THE BEHAVIOUR OF GASES

'Boyle's law' says that the pressure of a gas is inversely proportional to its volume, at constant temperatures. So the less space the gas has to move about in, the more pressured it is. Which is kind of obvious in theory, but doesn't actually happen with perfect predictability in reality.

'Charles's law' turns this around, saying that at a constant pressure, the volume of a gas is directly proportional to its temperature. Heat things up and the gas volume increases. Like Boyle's law, it doesn't work perfectly in real life.

'Avogadro's hypothesis' says that two different gases of the same volume and at equal pressures and temperatures contain the same number of molecules. It's a hypothesis rather than a law for now, because we don't yet have a good way to count gas molecules.

LAWS OF MOTION

These equations look at the relationship between an object's acceleration, its starting velocity, its final velocity, the time taken from start to finish and the distance covered in the process. Let's call these 'a' (acceleration), 'u' (starting velocity), 'v' (final velocity), 't' (time) and 's' (distance). Acceleration is measured as the result of final velocity minus the starting velocity, divided by time.

$$a = \frac{v-u}{t}$$

Distance travelled is the time divided by the average speed, which you work out like this.

$$s = t \ \frac{(u+v)}{2}$$

WAVES, RADIATION AND SPACE

WAVES

A wave is a disturbance that carries energy through space and time. Some types of waves need to travel through matter of some sort (whether solid, liquid or gas), but others can travel through the vacuum of space.

Waves are easy to understand if we start with a kind that we can see. Take a trip to the seaside on a reasonably windy day and watch the waves as they roll in. It looks at first as if great volumes of water are moving from way out to sea towards the shore, but that's not the case. Hopefully there's a gull out on the sea or some other floating object. If so, watch it and you'll see that it moves up and down as the waves pass through its position, but it isn't carried forwards. The water isn't the wave, it's just the material through which the wave passes, and each water molecule just moves a short distance up and down. A Mexican wave going round a stadium is another good illustration of how waves work.

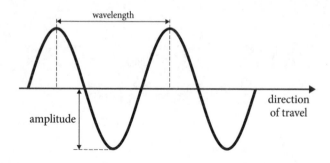

So, a wave is a pulse or vibration of energy that travels through space and/or through a conducting material of some kind. Waves come in two kinds – transverse and longitudinal. With transverse waves, the direction of the vibration is at right angles to the direction the wave travels. With longitudinal waves, the vibration and the wave move in the same direction.

To see what a transverse wave looks like, hold a long ribbon or a piece of string by one of its ends and flick it – you'll see the wave travel along to the other end. To see a longitudinal wave, you'll need to dig out your old Slinky toy and lie it flat on the ground. Pull one end out and push it in, and you'll see the wave travel along, the vibration moving at the same angle as the wave itself.

MEASURING WAVES

There are three measures you can take from a wave – its

amplitude, wavelength and frequency. The amplitude is how far the wave deviates from its undisturbed path, and the wavelength is the distance from one peak or trough to the next. This diagram shows what this means in a transverse wave.

Frequency is the number of waves produced per second. Its unit of measurement is the hertz. If you know the frequency and wavelength you can calculate the wave speed using this formula.

wave speed (metre per second) =
frequency (hertz) x wavelength (metre)

Sound waves (and also some types of seismic waves, the kind caused by earthquakes) are longitudinal waves. They can't travel through the vacuum of space but need a conducting material through which to pass, whether it's a solid, liquid or gas. The amplitude of a sound wave determines how loud the sound is (more amplitude = more volume, hence the term 'amplify'). A higher frequency produces a higher-pitched sound. If you keep turning up the frequency you will end up with sounds too high for human ears to detect, but they are still audible to some other animals.

Beyond 20,000 hertz (or, as it's more usually written, 20 kilohertz or 20kHz), we get into the range of ultrasound waves. We can't hear these, but their echoes can be used to form images of objects we can't see, like unborn babies.

Light is an example of a transverse wave. Light and other related wavelengths all fall along the electromagnetic spectrum, which is what we look at next.

THE ELECTROMAGNETIC SPECTRUM

We have already looked at the electromagnetic force. Now it's time to consider the electromagnetic spectrum, which contains all the electromagnetic wavelengths from radio waves at the low-frequency end to gamma radiation at the high-frequency end.

ELECTROMAGNO-WHAT?

The electromagnetic force, as we now know, is the one that's concerned with attraction between charged particles. This force in action also emits – or radiates – waves. The scientist Hertz discovered in the late nineteenth century that an electric current generates electromagnetic radiation, and this discovery led to the invention of radio.

In the twentieth century, scientists investigating the nature of electromagnetic waves discovered that they can behave either as waves or as particles called photons. This is the founding principle of quantum mechanics, which explores the relationship between the nature of an electromagnetic wave, and the amount of energy carried by its photons. The equation to express this relationship is this*.

$$e = h \times f$$

Here, e = energy, f = frequency and h is a conversion factor constant (Planck's Constant†).

---

*There are, of course, lots of other, much more complex equations associated with quantum mechanics, but this is as far as we're going to go with this seriously hardcore topic.

---

†The physicist Max Planck discovered that, at a subatomic level, energy can only be transferred in small, discrete units of the same quantity each time. The value of one of these units (or quanta) is 6.626176 x 10–24 joule-seconds, and is called Planck's Constant, in fact). Higher frequency waves carry more energy. Gamma radiation is used to kill cancer cells – that's how fiercely energetic it is.

Components of the Spectrum

We give different names to different sections of the spectrum because we have varied uses for them, but it's important to realize that electromagnetic radiation is a continuum, the difference between its components just a steady increase in frequency. There's nothing intrinsically special about the wavelengths of visible light either, it's just the bit we humans happen to be able to see, but other animals can see wavelengths that we cannot. Since photons are involved with the entire spectrum, it wouldn't be misleading to describe the whole lot as different wavelengths of light. Anyway, here's a simple diagram of how it works.

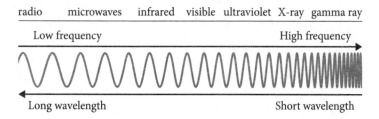

radio   microwaves   infrared   visible   ultraviolet X-ray gamma ray

Low frequency   High frequency

Long wavelength   Short wavelength

Categories of Frequencies

Radio waves occupy the low end of the spectrum, with frequencies up to about 10^4*. They are used to carry radio and television signals.

---

*i.e. 10 x 10 x 10 x 10 Hz.

Microwaves come next, their frequencies spanning 10^8–10^{12} Hz. We use them for cooking – they agitate water molecules and most of the stuff we eat contains a fair bit of water, so confining the food in a sealed chamber and bombarding it with microwaves is an effective way of making it hot. We also use them (at very low power) to carry Bluetooth and Wi-fi signals.

Infrared waves overlap with microwaves, covering roughly the 10^{11}–10^{14} Hz range. We experience them as heat, and we can use heat-sensitive imaging materials to generate 'heat pictures'. You might have seen them in action on exciting police chase documentaries, where the infrared cameras can pick out the fleeing criminal as a person-shaped glowing heat-source hiding in someone's garden at night.

Visible light comes next, occupying a small band around the 10^{15} mark. 'Pure' light combining all visible frequencies is white, but the objects around us absorb some of the wavelengths and reflect others, producing the range of colours we see around us. Shining a beam of white light through a prism onto a sheet of white paper bends the beam and separates out the visible frequencies – we can then see that the colours of light run through red, orange, yellow, green, blue and violet*.

---

*What happened to indigo? The transitions between the colours are arbitrary and subjective, and indigo is often left out. Pity. The same effect is produced when sunlight is refracted through water droplets on days with simultaneous sunshine and showers, producing a rainbow.

Ultraviolet light follows visible light, taking us up to about 10^{17}, though of course ultraviolet light is visible to plenty of animals. Naturally enough, it follows visible violet light on the spectrum. We know it as the thing that gives us suntans (and, unfortunately, sunburn and potentially skin cancer).

X-rays are around the 10^{18} mark. We know them as the means by which we can make images of stuff that's underneath other stuff, especially our bones. X-rays pass through some materials better than others, which is why they are good for looking at bones but not so good at helping us 'see' internal organs. X-rays can damage and kill cells, which is why the dentist leaves the room before x-raying your teeth – occasional exposure to x-rays won't do you any harm, but lots every day will do you no good at all.

Gamma rays have the highest frequency of all at 10^{19} or so, and they are responsible for the radiation we're talking about when we talk about radiation poisoning from spending too much time playing with uranium. Nuclear decay (when the nuclei of atoms break down to make simpler atoms) releases gamma radiation and it's going on in the universe at a tremendous rate, so there is a lot of gamma radiation out there – luckily our atmosphere absorbs it (though a huge burst of it at close quarters would certainly cause devastation). It can be very harmful to human tissues – although this can sometimes be used to positive effect, for example, as previously mentioned, it's used to kill cancer cells. Shielding made from very dense materials like lead is the best way to keep it out.

RADIOACTIVE SUBSTANCES

Naturally occurring radioactive elements are scarce on earth, but can be made in labs. They are elements or forms of elements in which the nuclei have an unstable combination of particles, so they spontaneously break down to form atoms of other simpler and more stable elements – this process is nuclear decay. Chemical elements towards the bottom of the periodic table are the most likely to be radioactive.

WHAT HAPPENS IN NUCLEAR DECAY?

Here's one of the (many) points where two science disciplines are intimately linked, in this case physics and chemistry. We'll be looking at the structure of atoms in detail in the topic 'Atomic structure' in the Chemistry section, so here we'll be skipping over those aspects rather quickly – for now just remember that an atom has a positively charged nucleus, surrounded by negatively charged electrons. When a nucleus decays, it releases one or more of three things – an alpha particle, a beta particle or a gamma ray.

The alpha particle is relatively big and heavy, and carries a positive charge. The beta particle is tiny and light and has a negative charge. Both of these particles can damage human tissues but they are relatively easy to block. The gamma ray has no mass and no charge, but unlike the other two it has a very high penetrative power and gets through most materials. Alpha and gamma radiation are the most dangerous kinds.

Discovery of Radioactivity

Most of us nowadays associate the scientist Marie Curie with the cancer care charity that bears her name. In the late 1890s she was investigating the rays emitted by uranium salts when she determined that the radiation was coming directly from the substances' atoms rather than through chemical reactions between atoms. She coined the term radioactivity to describe the effect she discovered, and also discovered two new elements, polonium and radium.

Unfortunately, she and her physicist husband, Pierre Curie, both paid a high price for their work,

HALF-LIFE

This term describes how long it takes half the atoms of a given quantity of a radioactive substance to undergo nuclear decay. Because we know that the radioactive form of carbon called carbon-14 (which decays to form nitrogen) has a half-life of 5,700 years, we can calculate the age of a fossil by seeing how much carbon-14 it has compared to its amount of carbon-12 – a non-radioactive form of carbon*.

---

*Carbon-14 and carbon-12 are examples of isotopes, of which there's more in the Chemistry section, and they nicely illustrate the point that 'isotope' doesn't always equate to 'radioactive'.

suffering from radiation poisoning because they were unaware at the time how harmful radiation was and had no protection from any of the products of radioactive decay. Marie died in 1934 (her husband had already been killed in an accident) from acute radiation sickness, and even now all of her research notes are too radioactive for safe handling and are kept in lead-lined storage boxes.

SAFE HANDLING OF RADIOACTIVE MATERIALS

Now we know how dangerous radiation is, we won't be in any hurry to start juggling chunks of uranium without some sort of special protection. Lead is dense enough to significantly cut down gamma radiation, so it's often used as shielding in labs where radioactive materials are used. Fallout shelters, designed to protect people from radiation after a nuclear bomb attack, are shielded with various materials – if they are underground, a metre of packed earth on the roof is necessary to block the gamma rays.

Safely disposing of nuclear waste from nuclear power stations is a real problem, as the spent nuclear fuel is still somewhat radioactive. It needs to be treated to make it as inert as possible, and sometimes stored deep underground, as isolated as possible from where people are living.

CHEMISTRY

THE PERIODIC TABLE

HOW THE TABLE WORKS

The periodic table lists all known chemical elements in order of their atomic number (the number of protons in their atoms). The table is displayed in such a way that elements with similar and recurring (or periodic) properties form columns. This is convenient for understanding periodic properties, but does produce a shape with large gaps between the left and right sides of the table in the first three rows. The ten columns that unite the two sides below the first three rows contain the transition elements, which bridge the gap between metals and non-metals. Two rows from the lower section (the lanthanides and the actinides) are usually extracted and shown separately at the bottom, as they have properties in common as you read across, rather than down.

A chemical element is a substance that is pure – in other words it can't be broken down by any chemical process into another substance, and all of its atoms have the same structure. Most elements are naturally unstable in the prevailing conditions here on planet

The Periodic Table of the Elements

1	2	3	4	5	6	7	8	9	10	11	12	13	14	15	16	17	18
1 **H** Hydrogen 1.00794																	2 **He** Helium 4.003
3 **Li** Lithium 6.941	4 **Be** Beryllium 9.012182											5 **B** Boron 10.811	6 **C** Carbon 12.0107	7 **N** Nitrogen 14.00674	8 **O** Oxygen 15.9994	9 **F** Fluorine 18.9984032	10 **Ne** Neon 20.1797
11 **Na** Sodium 23	12 **Mg** Magnesium 24											13 **Al** Aluminium 26.981538	14 **Si** Silicon 28.0855	15 **P** Phosphorus 30.973761	16 **S** Sulphur 32.066	17 **Cl** Chlorine 35.4527	18 **Ar** Argon 39.948
19 **K** Potassium 39.0983	20 **Ca** Calcium 40.078	21 **Sc** Scandium 44.955910	22 **Ti** Titanium 47.867	23 **V** Vanadium 50.9415	24 **Cr** Chromium 51.9961	25 **Mn** Manganese 54.938049	26 **Fe** Iron 55.845	27 **Co** Cobalt 58.933200	28 **Ni** Nickel 58.6934	29 **Cu** Copper 63.546	30 **Zn** Zinc 65.39	31 **Ga** Gallium 69.723	32 **Ge** Germanium 72.61	33 **As** Arsenic 74.92160	34 **Se** Selenium 78.96	35 **Br** Bromine 79.904	36 **Kr** Krypton 83.80
37 **Rb** Rubidium 85.4678	38 **Sr** Strontium 87.62	39 **Y** Yttrium 88.90585	40 **Zr** Zirconium 91.224	41 **Nb** Niobium 92.90638	42 **Mo** Molybdenum 95.94	43 **Tc** Technetium (98)	44 **Ru** Ruthenium 101.07	45 **Rh** Rhodium 102.90550	46 **Pd** Palladium 106.42	47 **Ag** Silver 107.8682	48 **Cd** Cadmium 112.411	49 **In** Indium 114.818	50 **Sn** Tin 118.710	51 **Sb** Antimony 121.760	52 **Te** Tellurium 127.60	53 **I** Iodine 126.90447	54 **Xe** Xenon 131.29
55 **Cs** Cesium 132.90545	56 **Ba** Barium 137.327	57 **La** Lanthanum 138.9055	72 **Hf** Hafnium 178.49	73 **Ta** Tantalum 180.9479	74 **W** Tungsten 183.84	75 **Re** Rhenium 186.207	76 **Os** Osmium 190.23	77 **Ir** Iridium 192.217	78 **Pt** Platinum 195.078	79 **Au** Gold 196.96655	80 **Hg** Mercury 200.59	81 **Tl** Thallium 204.3833	82 **Pb** Lead 207.2	83 **Bi** Bismuth 208.98038	84 **Po** Polonium (209)	85 **At** Astatine (210)	86 **Rn** Radon (222)
87 **Fr** Francium (223)	88 **Ra** Radium (226)	89 **Ac** Actinium (227)	104 **Rf** Rutherfordium (267)	105 **Db** Dubnium (268)	106 **Sg** Seaborgium (271)	107 **Bh** Bohrium (272)	108 **Hs** Hassium (270)	109 **Mt** Meitnerium (276)	110 **Ds** Darmstadtium (281)	111 **Rg** Roentgenium (280)	112 **Uub** Ununbium (285)	113 **Uut** Ununtrium (284)	114 **Uuq** Ununquadium (289)	115 **Uup** Ununpentium (288)	116 **Uuh** Ununhexium (293)	117 **Uus** Ununseptium (294)	118 **Uuo** Ununoctium (294)

58 **Ce** Cerium 140.116	59 **Pr** Praseodymium 140.90765	60 **Nd** Neodymium 144.24	61 **Pm** Promethium (145)	62 **Sm** Samarium 150.36	63 **Eu** Europium 151.964	64 **Gd** Gadolinium 157.25	65 **Tb** Terbium 158.92534	66 **Dy** Dysprosium 162.50	67 **Ho** Holmium 164.93032	68 **Er** Erbium 167.26	69 **Tm** Thulium 168.93421	70 **Yb** Ytterbium 173.04	71 **Lu** Lutetium 174.967
90 **Th** Thorium 232.0381	91 **Pa** Protactinium 231.03588	92 **U** Uranium 238.0289	93 **Np** Neptunium (237)	94 **Pu** Plutonium (244)	95 **Am** Americium (243)	96 **Cm** Curium (247)	97 **Bk** Berkelium (247)	98 **Cf** Californium (251)	99 **Es** Einsteinium (252)	100 **Fm** Fermium (257)	101 **Md** Mendelevium (258)	102 **No** Nobelium (259)	103 **Lr** Lawrencium (262)

CHEMICAL NICKNAMES

Each element has a shortened name, which is often
used in displays of the periodic table and always
in chemistry equations. Most of them are fairly
obvious shortenings of the full name, like H for
hydrogen, He for helium, Al for aluminium and so
on. Where this is not the case, it is generally because
there is another language involved. For example,
Au for gold comes from the Latin word aurum,
meaning, well, 'gold'. Na for sodium also comes from
its Latin name (natrium). All the nicknames are one
or two letters long apart from those for Ununbium
(Uub), Ununhexium (Uuh), Ununoctium (Uuo),
Ununpentium (Uup), Ununquadium (Uuq),
Ununseptium (Uus) and Ununtrium (Uut), and
frankly the less said about them the better.

Earth, and inclined to react with each other to form
stable compounds that are nothing like their elemental
components. For example, you probably won't encounter
chlorine, a smelly and poisonous green gas, or sodium, a
squishy metal that fizzes furiously in contact with water,

in their natural state anywhere outside a chemistry laboratory. However, you will find the stable compound formed between them – sodium chloride aka 'salt' – in your kitchen cupboard and many other places besides*.

When the Russian scientist Dimitri Mendeleyev was first putting together the periodic table there were many gaps, but the predictable repeated patterns of element properties meant that he could spot where an element was missing. This also meant that it was possible to make a very good guess as to what the missing element would be like and how to find it – and that process continues today. No element above the atomic number 92 exists in nature (on this planet, anyway) but they have been created in particle accelerators.

SOME ELEMENT GROUPS

Columns of elements in the periodic table that show periodic properties are called groups. Let's take a look at some of the most distinctive groups.

Alkali metals, the first group, are soft and very reactive metals that oxidize (react with oxygen) quickly. In the lab, they are normally kept in oil, until it's time to cut a

---

*You'll find out later that lots of things are 'salt' in chemistry, but not all of them are delicious on chips.

piece off and throw it into a dish of water to demonstrate how reactive it is. The reaction becomes increasingly violent as you move down the group, with lithium at the top fizzing around gently, and caesium near the bottom producing a violent explosion.

Alkali earth metals comprise the second group. They too are soft and reactive, though less so than the alkali metals.

Halogens, which most people may have heard of due to halogen lamps or bulbs, start out at the top of the column as gases (fluorine, chlorine) then liquid (bromine) and finally solids (iodine, and astatine – the rarest naturally occuring element with about 30 grams in the entire crust of the Earth). Halogens become less reactive as you move down the group, in contrast to alkali metals, but all are dangerous in their pure form. Fluorine is especially nasty* – most non-metals burst into flame in its presence, and it reacts even with something as innocuous as glass.

Noble gases are the very epitome of inertness, reacting with almost nothing. This is because each electron shell around their atoms has exactly the right number of electrons (see 'Chemical bonds' on p 67 for more about electron shells). They include helium, neon and argon, and while all are gases they get progressively heavier as you move down the group. A balloon filled with radon will plummet like a stone.

---

* We do however put it in our water in its reduced form, fluoride, to improve dental health.

ATOMIC STRUCTURE

A single atom is the smallest amount of an element that you can get – if you look at the constituent parts of an atom you'll end up with something else (and, probably, various other unrelated problems). Each atom is made up of equal numbers of positively charged protons and negatively charged electrons. Therefore the atom as a whole has no electric charge. Two or more atoms (whether of the same element or two or more different ones) bonded together become a molecule.

The protons form a central nucleus, while the much smaller electrons orbit this nucleus, in 'shells' or layers progressively further out from the nucleus, with each shell having an increased 'maximum capacity' of electrons. The nucleus contains variable numbers of a third particle type, the neutron, which has no electric charge. There are usually as many neutrons as there are protons, but not always … and the smallest atom of all, hydrogen, usually has no neutrons*.

Overleaf is a diagram of a carbon atom, showing how the electrons are arranged in their shells around the nucleus.

---

*We have talked about this in the Physics section (see p 33). It just goes to prove that you can't have chemistry without fully functional physics.

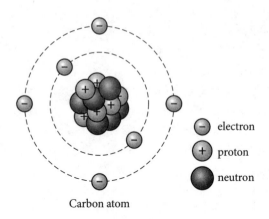

electron
proton
neutron

Carbon atom

An element's atomic number (the number that appears above its abbreviated name in most periodic table layouts) tells you how many protons it has in each atom (and therefore, by extension, how many electrons – under normal circumstances at least).

Atomic number and atomic mass

In some periodic tables you'll see two numbers – the atomic number (which progresses in single whole numbers from 1 for hydrogen, 2 for helium, 3 for lithium and so on), and the atomic mass, which is usually about double that of the atomic number at the top of the table but is getting on for three times the atomic number down at the bottom. The atomic mass isn't always shown as a

ANIONS AND CATIONS

If an atom has more or fewer electrons than it has protons, then it is an ion, with an electric charge. More electrons than protons = negative charge = anion. Fewer electrons than protons = positive charge = cation. Molecules with missing or extra electrons are ions too.

So how might an atom lose or gain an electron? Usually this happens during chemical reactions (of which more later). Here on earth most ions don't hang around long in normal outdoor environments as they are inherently unstable; but we can make them artificially by applying a lot of the right kind of radiation in controlled conditions. We also make them without thinking about it inside our own bodies, where they do vital and useful things like allowing our nerve cells to send electrical impulses. Ions are important in many natural and industrial processes, and stars are made of ionized gas (plasma).

All of which gives rise to the best (if not the only) chemistry joke of all time. Two hydrogen atoms are walking down the road, when one stops dead with a look of great anxiety. The other says, 'Hey, what's wrong?' The first says 'I lost my electron!' The second says 'Oh, no! Are you sure?' The first replies, 'Yes. I'm positive.'

whole number, e.g. in the table on p 59, for fluorine it's 18.9884032.

We already know that the atomic number of an element denotes the number of protons in one atom. Atomic mass can be considered as the number of protons plus neutrons. Actually, the more accurate definition is a fair bit more complex than that, but for our purposes this definition works quite well.

Obviously, for an individual atom, the atomic mass will always be a whole number. A carbon atom contains six protons and six neutrons, so its atomic mass is 12. But wait, not every carbon atom has six neutrons; most do – in fact 98.89% do in natural conditions. But the others have either one or two extra neutrons. The three varieties are called isotopes. They are specified by their atomic mass – the normal carbon isotope with an atomic mass of 12 is called carbon-12, the isotope with one extra neutron is carbon-13, and the one with two extras is carbon-14. The first two are stable but carbon-14 is not – it's formed in the atmosphere by cosmic radiation, but is radioactive and turns quickly into nitrogen.

The reason, therefore, that these values in periodic tables are not whole numbers is that they proportionately express the average atomic weight of all the known isotopes – this average is called the atomic weight, or relative atomic mass. In carbon's case, that's 12.0107.

Even if you have limited knowledge about nuclear

power, you'll probably put 'isotopes', 'radioactivity' and 'nuclear power' in the same part of your memory. Radioactivity and nuclear reactions happen because of the behaviour of the neutrons within atoms of unstable isotopes – there's more about this in the Physics section (see p 53).

CHEMICAL BONDS

Remember the carbon atom from the previous section? It has six of everything – six neutrons and six protons in the nucleus, and six electrons orbiting that nucleus. The electrons are arranged in an inner shell of two, and an outer shell of four.

So far, so good. But every electron shell has a maximum number of electrons that it can accommodate, and the second shell can hold up to eight of them. In the case of a carbon atom there are 'vacancies' in that outer shell for four more electrons. Therefore each carbon atom is unstable on its own and 'wants' to combine with other atoms – whether they are of carbon or a different element – to form a molecule whose constituent atoms each have a 'full' outer shell containing eight electrons, some of which are shared between the atoms. Such combining of atoms is a chemical reaction.

HYDROGEN AND HELIUM

The simplest example of how this works comes from the simplest element – hydrogen. It has only one electron, but it 'wants' to have two, because the innermost electron shell is only complete with two. Therefore, hydrogen in its natural state comes in molecules of two atoms (H_2), which share their two electrons so each has a full shell. They can be represented like this:

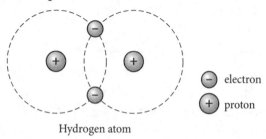

electron
proton

Hydrogen atom

The next element along is helium, which has two electrons per atom. That completes the electron shell without the need for any combining. Therefore helium naturally exists as individual atoms, and is disinclined to react with other substances. All of the other noble gases have complete electron shells too, so are similarly unreactive. But most other elements have incomplete electron shells, and so in their natural state don't occur as individual atoms but as multi-atom molecules. Here's another example – the carbon dioxide molecule, formed from one carbon and two oxygen atoms. Each oxygen atom shares two of its outer-shell electrons with the carbon atom, and the carbon atom shares all four of its outer-shell electrons, so all three atoms have a full outer shell of eight electrons.

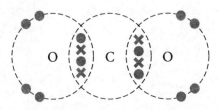

● electron in the oxygen atom's outer shell
✖ electron in the carbon atom's outer shell

TYPES OF BOND

When atoms share electrons like this, the bond formed is called a covalent bond. It is generally a very strong chemical bond, especially if two or more electrons are shared. Covalent bonds tend to be weaker when they involve electron shells that are further out from the nucleus. The valency of an element (the number of chemical bonds formed by its atoms) tells you how many electrons it 'needs' to make a compound. Hydrogen has a valency of 1, oxygen's is 2, therefore their compound – water – is H_2O.

Ionic bonds are very strong bonds formed between metals and non-metals, when each has either lost or gained an electron to become an ion (see Atomic structure p 63). While covalent bonds involve sharing electrons, ionic bonds involve exchanging them. Sodium and chlorine form an ionic bond when sodium loses its single outer shell atom to become a positively charged ion with two full electron shells, and chlorine takes up that electron to become a negatively charged ion with three full electron shells. In case you tuned out during

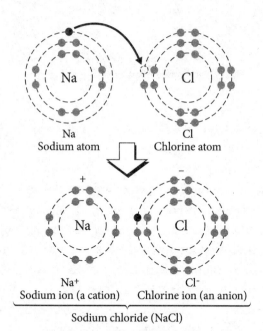

Na
Sodium atom

Cl
Chlorine atom

Na⁺
Sodium ion (a cation)

Cl⁻
Chlorine ion (an anion)

Sodium chloride (NaCl)

● electron
● electron moving to the chlorine atom
○ where the removed electron attaches itself, creating ionic bond

that last sentence, hopefully this illustration will make things clearer.

There are several kinds of bonds between molecules. They are what keep a solid substance like ice, carbon or titanium held together, often in a regular, crystalline arrangement, but in general the bonds are much weaker than the ones that form between atoms. One of the most common is the hydrogen bond, which is formed by the attraction between the positively charged nucleus of a hydrogen atom within one molecule, and a different kind of atom in another molecule. It is hydrogen bonds that hold water molecules firmly together in ice, and loosely together in liquid water.

The metallic bond is what holds the atoms of a metal together. It's like a covalent bond in that electrons are shared, but they are shared between all atoms at once, producing what's sometimes called a 'sea of electrons' shared freely between the nuclei of the metal atoms – the structure is sometimes called a metallic matrix. Therefore, it makes more sense to refer to 'metallic bonding' as there is no single individual bond. This type of bonding is what makes metals ductile (stretchy) and/or malleable (squashy rather than brittle).

ELECTRONEGATIVITY

Sharing electrons sounds very fair-minded, but some kinds of atoms are less willing to share than others. Let's consider that most volatile of elements – fluorine. Fluorine atoms pull electrons towards themselves more strongly than any other kind – they have the highest electronegativity. So in a molecule of hydrogen fluoride, the two electrons that form the bond between them are always closer to the fluoride end, giving that end of the molecule a negative charge and the hydrogen end a positive charge. Molecules like this are called polar molecules.

Solids, Liquids and Gases

Bonds, whatever their type, are what determine the state or phase of a substance – whether it is solid, liquid or gas. Let's look at a nice ordinary substance, say, water. Fresh from the freezer it is a solid. Its atoms connect up into regular crystals, which are joined to each other in an equally regular pattern. In ice the crystals are like hexagonal wheels. In other solids they may be different in shape but all true solids have a regular crystalline structure of some kind or another. Those that don't, like glass, are really just very slow-flowing liquids.

After a while at room temperature (specifically, above 0°C) the ice crystals will melt. That energy is enough to start breaking down the hydrogen bonds that hold the structure together, so a phase change occurs from solid to liquid. Turn up the thermostat to a toasty 100°C (or, more practically, put your water in the kettle and switch it on) and it will boil – the hydrogen bonds all completely break down and the water becomes a free gas (or steam).

CHEMICAL REACTIONS

We've already seen why elements react together to form compounds – it's to do with going from an unstable state with incomplete electron shells to a stable one with complete shells. Compounds may also react together for the same reasons, because some compounds have a more stable structure than others. The chemical bonds that hold the atoms of a compound together are generally strong, and under normal circumstances aren't readily broken. For that reason we don't observe many rapid, dramatic chemical reactions going on in the world outside, although there are any number of them going on all the time inside us.

THE MATHEMATICS OF CHEMISTRY

Chemical reactions are usually shown as equations, with the reactants (things that are reacting) on the left, and the products (what they turn into) on the right, joined by an arrow. Here's an example:

$$CH_4 + 2\,O_2 \longrightarrow CO_2 + 2\,H_2O$$

This equation is what happens when you burn methane gas. Every molecule of methane (CH_4) reacts with two molecules of oxygen (O_2) to produce one molecule of carbon dioxide (CO_2) and two molecules of water (H_2O).

Equations involving charged atoms (ions) may show the charges on the atoms involved, for example, this is the reaction between silver (Ag) and chloride (Cl) ions.

$$Ag+ + Cl- = AgCl$$

Rust to rust

Any reaction between a metal and a gas is called oxidation, though the term is most commonly used (for obvious reasons) when the gas in question is oxygen.

They say a watched pot never boils. It's not true, it's just very boring waiting for it to happen. Even more boring is waiting for the pot to go rusty. We tend to think of chemical reactions as rapid and dramatic, but the reaction between iron and oxygen is slow and gradual. The resulting crumbly orange-red stuff is an iron oxide, a compound of iron and oxygen with ionic bonds. Water is necessary to make rusting occur because it acts as an electrolyte – helping the free electrons released from the iron atoms to unite with the oxygen atoms.

The process is indeed slow by our standards, but has been going on across planet Earth for a long time, which is why there's very little pure iron to be found naturally anywhere. Therefore it's only man-made iron things that we observe going rusty. So ... where did we get the iron from to make the pot, the bicycle and all that other rusty stuff? How do we turn iron oxide back into iron?

First, you need some iron oxide. This is found naturally in the form of iron ore, rocks rich in iron oxides. They tend to have a giveaway reddish colour. Heating up iron ore in the presence of carbon eventually causes the oxygen to separate from the iron and instead bond with the carbon, which it does readily as carbon-oxygen bonds are stronger than iron-oxygen ones. The result is the production of carbon dioxide and pure molten iron*.

ENERGY

As the iron-making example shows, you will often have to put in lots of energy to bring about a chemical reaction. When you take ice out of the freezer, the energy from its new warmer surroundings starts to break the hydrogen bonds that hold the molecules together, freeing the molecules to move around more, so it becomes liquid. More heat and more bonds are broken, until you end up with a gas in which the molecules aren't bonded together at all. It takes a whole lot more energy to crack the bonds between the atoms themselves.

Set fire to a piece of wood, and the heat starts to break down the various hydrocarbon compounds that it's made of, releasing hydrogen and carbon which reacts with oxygen in the air to produce smoke. By the time you've finished, what's left is a black lump of almost pure carbon.

So, it takes energy to break up compounds, but when

---

*As discovered in the Iron Age, when our ancestors cooked their mammoth steaks over hot stones and were surprised to observe liquid metal coming out of the stones.

ACID TEST

An acid is a compound which, when dissolved in water, produces hydrogen ions. An alkali (or, to give it its more correct name, a base) is something that, when dissolved, accepts hydrogen ions. The pH (potential of hydrogen) scale expresses how acidic something is – pure water has a pH of 7, which is the mid-point, acids come in at under 7 and bases over 7. When you put an acid and a base together, you get a reaction that produces a salt of some kind, with water as a by-product. One way of making everyone's favourite salt, sodium chloride, is to combine dissolved sodium hydroxide with hydrocholoric acid. Then you just get it warm enough for the water to evaporate, and you'll be left with solid sodium chloride crystals.

compounds are formed, energy is released. Think back to that lump of sodium in the dish of water. As the sodium combines with the hydrogen and oxygen in the water to produce sodium hydroxide, lots of energy is generated, making the sodium lump fizz and bounce around on the water's surface. If your sodium lump is too big, you'll get an explosion.

COLLISION THEORY AND RATES OF REACTION

So, we know that some chemicals react with others, under the right conditions. We can also observe that the reaction between the same two chemicals may be faster or slower, depending on a variety of external factors. To explain exactly how reactions occur and why their speed may vary, chemists Max Trautz and William Lewis came up with collision theory.

The key premise to collision theory is simple – to react, two molecules or atoms must literally collide, with sufficient speed and at the right angle that existing chemical bonds are broken. Suitable collisions will

CONCENTRATION SPAN

When you throw a handful of table salt in some water, the salt dissolves and you get salty water – a solution of sodium chloride. The sodium chloride doesn't actually react with the water to form a different compound, but its crystals completely break down into individual sodium and chloride ions, forming a homogenous (uniform) mixture. In this case the salt is the solute (the thing that dissolves) and the water is the solvent (the thing into which it dissolves).

The concentration of a solution is defined as the number of moles of solute per litre of the solution. Why the sudden mention of moles? Mole is a unit of quantity in chemistry. One mole of anything is the amount that contains the same number of particles as there are in 12 grams of carbon of the carbon-12 isotope. Why this figure, and why moles*, is anyone's guess. We think of solutions as being solids dissolved in liquids, but you can also have solutions of gases dissolved in other gases (the air we breathe is an example – various gases dissolved in nitrogen), solids dissolved in other solids (as with stainless steel, in which carbon atoms are mixed in with the crystals of iron) or other combinations. The relative amounts of solute and solvent determine the concentration of the solution. The spread of molecules of solute through the solvent is called diffusion. The tendency for solute molecules to move from areas of high concentration to areas of low concentration is called osmosis.

---

*'Mole' is a versatile word. Small furry burrowing mammal, dark spot on the skin, undercover agent, unit of measurement in chemistry, a pier or breakwater, and also a big digging machine!

happen more often when the chemicals are at a higher concentration, and also under higher temperatures when the atoms/molecules are moving about more rapidly.

ANGLES OF INTEREST

If the molecules don't collide at the right angle, there won't be a reaction. Bonds need to be broken for a reaction to occur, and bonds (being formed by shared electrons) have a negative charge, so it takes a positive charge to break them. In the last section we looked at the molecule hydrogen fluoride, which is negatively charged at its fluoride end and positively charged at the hydrogen end. To break a bond in another molecule and cause a reaction, the hydrogen end of the molecule has to hit the bond. If the fluoride end hits, it's negative-meets-negative and the two repel each other.

REACTION RATES

A gentle collision doesn't break anything. A certain amount of energy is needed before anything interesting will happen – what is known as the activation energy. How much this is varies from reaction to reaction, but once it's been achieved then a reaction will start to occur.

How quickly the reaction happens now is determined by the concentration of the two things reacting together and affected by various other factors, and can be predicted by horribly complex-looking equations.

Catalysts

Catalyst is a chemistry term which has found its way into general parlance without really changing its meaning. In chemistry a catalyst is something that causes a reaction rate to increase – it 'helps things along' if you like. Therefore, catalysts are much used in industry to speed up chemical reactions. Usually what happens is that the catalyst first reacts readily with one of the reactants to form a compound that then combines with the other reactant. The unstable result then quickly turns into the expected product plus the catalyst reconstituted. We can show this as a sequence – let's say A is the first reactant, B the second reactant, P the product of a reaction between them and C a catalyst.

{*without catalyst*} $A + B \longrightarrow P$

{*with catalyst*} $A + C \longrightarrow AC$
$$AC + B \longrightarrow ABC$$
$$ABC \longrightarrow CP$$
$$CP \longrightarrow C + P$$

The 'with catalyst' route looks more convoluted but because all the stages happen much more quickly than $A + B \longrightarrow P$, the overall reaction rate is increased. A huge variety of substances can work as catalysts under a number of different conditions.

FUELS, AIR, POLLUTION

CHEMICALS IN THE AIR

Here on planet Earth, the air we breathe is a mixture of various gases, and by volume it's a shade over 78% nitrogen. Nitrogen forms molecules of two atoms each, joined by a triple covalent bond (formed of six electrons, three from each atom). This bond is extremely strong, making N_2 very unreactive. That's good news for us, as it means we can breathe it in and out without it reacting with anything inside our bodies, and it doesn't readily react with other things that we might accidentally or on purpose let loose into Earth's atmosphere.

The next most abundant gas in our air is oxygen, making up nearly 21%. Oxygen also forms two-atom molecules, but with a single rather than a double bond, making oxygen more reactive. Animals consume oxygen, but luckily plants produce it.

THE FINAL 1%

The remaining 1% of air is composed mostly of argon, one of the noble or inert gases. The other noble gases put in an appearance too, with helium, neon, krypton and xenon contributing trace amounts. Carbon dioxide makes up about 0.03% and there are tiny amounts of hydrogen, methane, carbon monoxide and assorted others.

Water vapour is usually excluded from breakdowns of air composition, because its presence is extremely variable, given its tendency to turn into rain or snow and fall down, only to evaporate under the sun and rise up again. It can make up as much as 4% of the atmosphere (air that is 4% water vapour is at 100% humidity).

AIR POLLUTION

So much for what *should* be in our atmosphere – what about the harmful stuff that shouldn't, and is causing climate change and other mayhem? Air pollutants are harmful (to people and/or the environment) substances of any kind at large in the air. Most of them have been put there as a result of human activity, though some come from natural sources, such as volcanoes or animal farts. Some are gases, while others are particulate matter (tiny bits of solid or liquid suspended in the air). The most significant baddies are as follows ...

Sulphur dioxide and other oxides of sulphur, produced by burning fossil fuels and other industrial processes (and volcanoes). Responsible for acid rain.

Nitrogen dioxide and other oxides of nitrogen, produced by burning substances at very high temperatures. A key component of smog.

Carbon dioxide and carbon monoxide, produced by burning carbon-rich material. Carbon dioxide is an essential atmospheric gas in the right quantities as plants require it, but in excess it works as a greenhouse gas, preventing the sun's radiation from exiting the atmosphere and thus increasing global warming. Carbon monoxide is extremely poisonous.

Methane and similar hydrocarbon (made of hydrogen and carbon – also referred to as organic) compounds are poisonous and may act as greenhouse gases.

Particulate matter covers a vast array of substances, from asbestos fibres to sea salt, and rock dust to carbon. Most of it you don't want to inhale, and it can have dramatically bad environmental effects too.

THE OZONE PARADOX

Ozone is an unstable form of oxygen in which each molecule holds three atoms instead of two – O_3. While its presence in the upper atmosphere is vital to protect us from ultraviolet radiation, at ground level it can be a dangerous pollutant. It is formed by sunlight acting on molecules of other air pollutants like nitrogen oxides. The ozone layer has been thinning out by about 4% a decade since the 1970s, with more dramatic thinning at both of the poles – mainly because of increased chlorofluorocarbon (CFC) compounds, which we use as refrigerants, solvents and various other things. The CFC molecules lose their chloride and fluoride ions when they reach the upper atmosphere and get battered by ultraviolet rays. These ions then react with the ozone molecules.

MEASURING POLLUTANTS

The amount of a pollutant in the air is measured by parts per million by volume (ppmv). Wait, what? What's a part, a million whats? It depends what you're measuring, but if it's a gaseous pollutant in air, you're going to be talking about millilitres of pollutant per million millilitres of air. With particles, the measurement is more likely to be micrograms or milligrams per cubic metre of air.

METHODS OF MEASUREMENT

There are several different ways to measure pollutants in the air. Passive sampling is the simplest – a local authority worker just unseals one end of their diffusion tube, places it with the open end downwards wherever they want to measure air quality, leaves it there for a month, then sends it off to their friendly local laboratory. The tubes contain steel gauze coated with a chemical, which reacts with the pollutant the local authority worker is interested in. Usually diffusion tubes are used to detect nitrogen dioxides, and the chemical used is triethanolamine, which converts the NO_2 to nitrites, which the lab analyzes.

Active sampling involves taking a known quantity of air and testing it for a variety of chemical pollutants. Most air quality monitoring sites carry out both passive and active monitoring. The active kind is more likely to be used to monitor pollution* around specific locations where there's cause for concern at certain times.

Researchers also measure water and soil quality for pollutants, both routinely and in response to natural or human-caused disasters. Drinking water, in particular, is tested for the presence of nasties like heavy metals, dissolved salts, pesticides and many other substances.

USEFUL CHEMICALS FROM CRUDE OIL

It's a precious natural resource, over which wars are fought and territory disputed, and it's running out fast. Crude oil is a fossil fuel, formed from billions of long-dead ancient algae and microscopic water animals which sank down to lake bottoms or seabeds and were then mixed up with mud and squashed under layers upon layers of sediment in an oxygen-free state. These conditions gradually turned the solid organic material into liquid and gas. A similar process with land plants

---

*You can check forecasts of air pollution, just as you can for the weather. In fact, temperature and airflow partly determine how bad air pollution will be for any given place and time.

was responsible for the production of coal.

Crude oil contains lots of different kinds of hydrocarbon molecules. Carbon and hydrogen tend to combine to form rings or long, sometimes branching chains (or combinations of both). Nearly all the vehicle fuels we humans use at the moment are derived from crude oil, though this is going to have to change sooner or later as this is not a renewable resource. Estimates of how long the oil we have left will last varies greatly but most put it at well under a hundred years, even allowing for the 'unproven' reserves that should be out there but that we haven't found yet*. Additionally, burning fossil fuels is a major source of pollutants.

FUELS AND MORE

Here are some of the most familiar everyday chemicals that are distilled from crude oil. Distillation is the process of separating out certain components from a liquid by boiling it and condensing back the vapour that is produced at particular temperatures.

Petrol fuels most of our cars and other motor vehicles. It is comprised mainly of fairly short-chain hydrocarbons,

---

*There is more coal left than there is oil, but that'll all be gone in a hundred and fifty years or so.

with between four and twelve carbon atoms per molecule. Not that long ago, petrol in most countries came with added lead, which helped protect engines but also, unfortunately, helped people get lead poisoning, so it is being phased out.

Diesel is a popular alternative to petrol, especially for larger motor vehicles as it gives you more mileage for your money. It contains longer-chain hydrocarbons than petrol, as well as those based on hydrocarbon rings. Jet fuel is chemically quite similar to diesel.

Kerosene or paraffin is widely used as a fuel for cooking and heating. It's also, more glamorously, burned with liquid oxygen as rocket fuel. It is composed mostly of six- to sixteen-chain hydrocarbons.

Lubricants for engines of various kinds, to protect moving parts from wear, are derived from crude oil. So are alkenes, which are used to make plastics, and asphalt, and tar.

MAKING LIFE CYCLE ASSESSMENTS

Assuming we don't annihilate ourselves or our planet any time soon, our descendents will probably – hopefully – be around to learn about the turn of the millennium as a troubled time for humanity, when the devastating effects of our industries on the earth were finally undeniable, and we were slowly, reluctantly and fearfully beginning to find new ways forwards.

Taking a realistic and detailed look at the impact of a given product on the earth, from the manufacture of its components from raw materials to its disposal at the end of its 'working' life, is an important first step to working out how we can reduce that impact. Making the results public means that we consumers can make informed choices about what products and services we use. The life cycle assessment (LCA) is a part of the ISO (International Organization for Standardization) 14,000 standards for environmental management in businesses of all kinds.

WHAT GETS ASSESSED?

Ideally, everything. In terms of a manufactured product, the extraction of raw materials, the water used in any purification or distillation process and the fuel burnt to transport those materials around all need to be measured. If the product is meat or veg, you'll need to assess how much water and other resources it consumes as it grows. The actual manufacturing process itself will use more resources and generate more pollution. Packaging, distribution to outlets for sale, the things the product uses up in its actual function, how recyclable the product is and the likelihood that it will actually end up being appropriately recycled ... all need to be assessed. The level of detail is also important – do you make a single calculation for road miles covered, or factor in relative amounts of congestion where the delivery vehicle can't be driven so economically*?

It's not surprising that coming up with simple figures for LCAs is difficult, and two assessments for the same product may produce quite different results. The science involved is in its infancy and no doubt we'll get better at it – in the first place, though, LCAs should help identify those products which have the most severe environmental impact, and point the way to making improvements.

---

*This is why you should buy locally grown, seasonal food. Strawberries aren't for Christmas, parsnips are.

METALS

THE EARTH'S STRUCTURE

Once we humans figured out that our planet was a sphere*
rather than a flat disc, we wondered what lay below the
visible surface, all the way down to the centre of the sphere.
No one's been in all the way to check, but the amount of
gravity the Earth exerts on other objects in space allows us
to calculate its density as a whole, which in turn helps us
to work out the composition of the layers that are deeper
than we can actually physically investigate. The conclusion
is that there are several main layers – solid on top (the
crust), becoming viscously flowing semi-solid below that
(the mantle), then a more fluid liquid layer (the outer core)
and finally the very inside (inner core), which is solid.

Elements of Earth

As the Earth formed, its denser materials would have
migrated to the centre. Iron, one of the most abundant
elements on earth, is thought to constitute about 80% of

---

*Actually it's an oblate spheroid – a slightly squashed sphere.

the inner core. It has been proposed by some scientists that the core is actually a single colossal crystal of pure iron, but most believe it is mainly iron with some nickel mixed in – nickel is a close neighbour to iron in the periodic table. The extreme pressure on the inner core means the iron stays solid at temperatures that would normally melt it.

The outer core is under less pressure but it, too, is screamingly hot, so the metal of which it is made (again, probably a mixture of mainly iron with a bit of nickel) is liquid. All this iron in the outer core is responsible for the magnetic field of Earth.

The mantle is the bit that sometimes makes an appearance via volcanoes. It is composed of molten rock, which is rich in silicon, magnesium and, once again, iron, as well as tiny amounts of a whole range of other elements. The iron and most other metals here are not pure, but held in oxides and other compounds. There is a continuous exchange going on between the molten rock of the mantle and the solid rock of the crust. Volcanoes release lava – molten rock from the mantle, which in due course cools down and becomes surface rock. This gets eroded by wind and water, and its dust is deposited elsewhere as sediment. The build-up of sediment means that the lower layers are compressed down over immensely long time-spans until the heat and pressure melts them back into the mantle. *Et voilà*, the rock cycle.

The eight most common elements in the Earth's crust by mass are oxygen (46.6%), silicon (27.7%), aluminium (8.1%), iron (5%), calcium (3.6%), sodium (2.8%), potassium (2.6%) and magnesium (2.1%). A further 84 elements make up the remaining 1.5% or so.

BURIED TREASURE

Why do we like gold, silver and platinum so much? One reason is that when we find them in their natural state, they are recognizably pure gold, silver and platinum rather than grotty-looking oxides. These metals are very unreactive, which means that they don't react with everyday things like oxygen but instead stay lovely and shiny and are ideal for decorating stuff, including our fingers, necks and earlobes. All three are sometimes described as 'noble metals', drawing comparison with the unreactive noble gases. Other noble metals include rhodium, palladium, iridium and, according to some, mercury, but they are not so pretty and shiny – and, in the case of mercury, since it is liquid at room temperature, rather messy to wear. They are all very rare in the earth's crust, making them even more valuable to us.

METALS AND ALLOYS

As we saw in the Chemical bonds section, metals have the special property of metallic bonds, which lends them many of the qualities that make them stand out from the other elements. As well as being flexible and shiny, they are also excellent conductors of electricity and heat. These attributes make metals very important and useful to us.

TYPES OF METALS

Most of the elements in the periodic table are metals. You'll find those that aren't metallic clustered away on the right-hand side of the table, and between the two is a diagonal dividing line of elements like silicon, which have some metallic traits and are known as metalloids. Groups 14-16 in the table all transition from non-metal to metal as you go down, via two metalloid elements.

You've probably heard the expression base metal before, perhaps in a derogatory sense – it is not a technically precise term, but covers a variety of metals that corrode (rust) quite readily. The so-called noble metals (as discussed in the previous section) don't.

Scientists have also synthesized some compounds which have metallic traits, including the silvery-grey sheen and ability to conduct electricity.

ALLOYS

If you look around your immediate surroundings you'll probably quickly find a variety of things that are clearly made of metal. However, it's a fair bet that none of them are made of a single, pure, elemental metal. Often, mixing two metals or a metal and a non-metal together to form an alloy produces a more useful product. Pure gold, for example, is very soft and easily knocked out of shape, so gold used in jewellery or coinage is typically an alloy of gold and one of the base metals, to give it a bit of extra hardness. Bronze* is an alloy of copper and tin, while brass combines copper and zinc.

Steel is a very widely used alloy of iron and carbon – adding a small percentage (up to 2%) of carbon significantly strengthens the material. However, throw in too much carbon and the material becomes brittle. Adding chromium to steel produces stainless steel, which does not rust.

An alloy is not a compound, as the components don't react together, but it is a solution, as the new element's atoms are completely mingled within the metallic matrix. In steel, the carbon atoms fit between the iron atoms without disturbing the repeated-cube structure assumed

---

*The Bronze Age came before the Iron Age – bronze is tougher and arguably better-looking than iron, but much rarer.

by the iron atoms. When the structure is placed under stress, the carbon atoms help 'fill the gaps' when cracks form in the metal matrix.

CONSTRUCTION MATERIALS

Our ancestors established the use of various metals for making all manner of tools. Alloys of various kinds are suitable for different types of construction, and using metals alongside non-metallic materials like stone and wood, with their range of different properties, has enabled us to undertake construction on a massive scale, dramatically changing both everyday human life and the whole topography of the planet.

PROPERTIES OF CONSTRUCTION MATERIALS

When you're building a large structure – say a bridge, or a house, you are going to want it to stay standing for a good while, coping with a whole range of everyday forces. A bridge across a river must be strong enough to bear the weight of constant traffic, and it must be resistant to corrosion and erosion from moving water. The house must be strong enough to support its floors and the furniture that stands on them, and it must keep the weather out and the warmth inside. Measurable properties of construction materials include compressive strength (resistance to being squashed), tensile strength (resistance to being stretched) density, porosity, conductivity of heat and electricity, fire resistance and sound insulation.

STONE

Buildings intended to have people living in them are usually made out of some kind of rock or stone, as most are strong, durable and have good insulation properties. The basic rock raw product is usually cut or formed into bricks or blocks which are easily stackable, and often fixed together with cement or another bonding substance. Most modern bricks and building blocks are made from fired (heat-dried) clay or set concrete rather than being cut from solid stone. This means the manufacturer can control the consistency of the material as well as the shape of the blocks, and incorporate air spaces if necessary.

WHAT KIND OF METAL?

Steel is the most popular metal in construction today. The proportion of carbon to steel determines its strength. We have already seen how mixing steel with chromium produces stainless steel, which resists corrosion more effectively, and is also attractively shiny. It is mostly used for small tools, such as cutlery and surgical instruments. Steel used in construction also needs to be as rust-proof as possible, and a different method is used to achieve this – galvanization. Steel is galvanized by putting it in a 460°C zinc bath. In contact with air, the zinc reacts with oxygen and then the resultant zinc oxide reacts with carbon dioxide. The result is a coating of zinc carbonate, a dull inert compound that prevents further reaction. Galvanized steel is a tough, strong and very durable material, suitable for all kinds of outdoor constructions.

METALS IN CONSTRUCTION

Metals in small amounts are found in pretty much every kind of large-scale construction. The right alloy can offer enough hardness and resilience to stand up to extremely pressured conditions. However, some of the other metallic properties, such as its high heat conduction, mean it's not suitable for everything.

Metal is a good choice for long bridges, because of its high strength-to-weight ratio and also its high tensile strength, a property lacking in concrete and other stone-based materials. For similar reasons, metal is used for roofing, especially on large buildings. It is easy to coat metals with reflective paints, which can have significant cooling effects when used to roof buildings in hot environments.

ORGANIC CHEMISTRY

NATURAL POLYMERS AND THEIR ROLES IN NATURE

There are certain 'sciencey' words that you will often hear in adverts for wrinkle-smoothing creams or stomach-taming yoghurt drinks, and one of them is polymer. There will no doubt be others later on, but for now let's define polymer – it's a large molecule with a repeated structure, forming a chain. The single 'links' in the chain are called monomers; join up lots of monomers and you produce a polymer. Here's an example – the monomer molecule ethylene can join up to form the polymer molecule polyethylene, when the double bond between ethylene's two carbon atoms is broken, leaving each free to form a new bond with another carbon atom.

Ethylene   Polymerization   Polyethylene

Most natural polymers look a lot like this – they are based around a long chain of connected carbon atoms, with hydrogen atoms coming off to the side. Hydrogen and carbon-based polymers are called organic, because of their prevalence and importance in living things, although we can also synthesize them artificially. If they contain just hydrogen and carbon they are hydrocarbons – we looked at lots of them in the section on crude oil. If they incorporate oxygen too they are carbohydrates – you may well have noticed by now that molecules incorporating oxygen often end with 'ate'. Polymers may also include atoms of many other elements.

WHAT ARE POLYMERS FOR?

Lots of things in nature are made of polymers. One of the most important is DNA, or deoxyribonucleic acid, the molecule that provides the 'instructions' for making all the different kinds of proteins in our bodies. DNA is a pretty complex polymer, or to be precise two pretty complex polymers, as each strand of DNA comprises two polymer chains, connected by a 'ladder' of hydrogen bonds and twisted into the familiar 'double helix' spiral. The monomers are nucleotides, each comprising a simple sugar, a phosphate group and one of four kinds of nucleobase – don't worry too much about what all this means as we'll come back to it in the Biology section. For now, the point is that the pattern is repeated, again and again, to form a chain of (theoretically) endless length.

So DNA is one kind of natural polymer. Another, rather simpler but equally useful one (if you happen to be a plant) is cellulose, which is the main constituent of plant cell walls. It is also the main source of dietary fibre for us, helping to protect us from unpleasant intestinal issues, so it's actually not just plants that consider it important. Cellulose is a carbohydrate, composed of carbon, hydrogen and oxygen, and its structure is expressed like this:

$$(C_6 H_{10} O_5)\, n$$

The little 'n' shows that the structure is repeated as many times as you like. As a diagram it looks like this:

The ring-shaped monomers that make up carbohydrate molecules like cellulose are simple sugars. Therefore, molecules like these are sometimes called polysaccharides – 'many sugars'. Another name for polysaccharides is starches. Our bodies break down starches like glycogen – $(C_6H_{10}O_5)n$ – to release simple sugars like glucose, which are then used in chemical reactions in our cells to provide energy.

POLYUNSATURATED?

Another kind of polymer you'll find in your cells is fatty acid. These molecules have lots of jobs within our bodies, and we eat them in the form of fats. You probably know that polyunsaturated = good and saturated = bad, but what's the difference? A saturated fatty acid is one in which all the carbon atoms (except the one at the end of the chain – the 'acid' part) are joined to two hydrogen atoms, and to each other by single bonds. If one or more adjacent pairs of carbon atoms have only one hydrogen atom each they are joined to each other by a double bond instead. Because there is 'room' for additional hydrogen atoms in the molecule, it is 'unsaturated' – monounsaturated if there is one double bond, polyunsaturated if there are several. No room for more hydrogen = saturated.

The essential fatty acids we all need to consume

```
      H   H   H   H   H   H   H   H   H        O
      |   |   |   |   |   |   |   |   |       //
  H — C — C — C — C — C — C — C — C — C — C
      |   |   |   |   |   |   |   |   |       \
      H   H   H   H   H   H   H   H   H        O
                                               |
                                               H
```

Saturated fatty acid

```
                                    H   H    O
                                    |   |   //
                                    C — C — C
                             H   H //        \
                             |   |/           O
                         H   C — C — C         |
                         |   |       |         H
                     H   C — C       H   H
                     |   |//
             H   H   C — C
             |   | //
   H — C — C — C
       |   |   |
       H   H   H
```

Unsaturated fatty acid

because our bodies can't make them are unsaturated.
However, we can make our own saturated fats so we
really don't need to eat them. They get stored as a
long-term energy source, and too much of them makes
us fat.

Trans fats are unsaturated fats in which the carbon
chains are arranged in straight lines rather than with
kinks at each double bond. They don't occur widely
in nature – most of the ones we eat are formed from
processed vegetable oils used in the delicious-fried-
junk-food industry. The innocent-sounding structural
difference makes trans fats extremely bad for you
because they are much harder for the body to break
down and digest. Such is the wonder of chemistry.

NUTRITION

We're made of chemicals, and we use up chemicals constantly in our innumerable bodily processes, so we need to eat chemicals in order to keep living. The most convenient way is to eat the bodies of other living (well, living until recently) things, which are made up of the same kinds of chemicals as us. Dietary chemicals are called nutrients.

A BALANCED DIET

We need to eat proteins, carbohydrates and fats. As we saw in the previous section, carbohydrates and fats are both made up of carbon, hydrogen and oxygen. Proteins, which contain nitrogen as well as carbon, hydrogen and oxygen, are used to build structural tissues but also form 'working' molecules like antibodies, enzymes and hormones*.

THE ESSENTIALS

Some of the chemicals we need in our body are things we can synthesize ourselves. Others we cannot, and we have to eat them instead. Of the twenty 'standard' amino acids we use to construct our proteins, eight are called 'essential' –

---

*Not all hormones are proteins, but lots of them are.

essential in that we have to include them in our diets. They are isoleucine, leucine, lysine, methionine, phenylalanine, threonine, tryptophan and valine. The disease kwashiorkor is caused by a deficiency of essential amino acids. Luckily for most of us, they are all found in commonly consumed foods like most kinds of meat, grains, pulses and eggs.

There are essential fatty acids too. You may be among the many who takes them in the form of supplements – they are not as widely available in 'normal' foods as the essential amino acids. They come in two kinds – omega-3 and omega-6. Linolenic acid is an example of an omega-3, while the obviously completely different linoleic acid (who names these things?) is an omega-6. What about omega-9? I hear you ask. It exists, but it's not normally essential to include in our diet, since most of us can manufacture it in our bodies. Those with compromised liver function sometimes can't, however, which is why supplements often contain all three kinds. To avoid the need for supplements, make sure your diet includes fish and/or various oily seeds including those of flax, sunflower and pumpkin.

VITAMINS, MINERALS AND FIBRE

Vitamins are specific compounds which are vital (albeit often in tiny amounts) for healthy life, and which we can't manufacture ourselves (or can't make in sufficient quantities) but have to eat. Even quite closely related animals may vary in terms of which vitamins they need. For example, most mammals, including primates, can make their own vitamin C, but we can't, and nor can guinea pigs (guinea pigs, however, seem to have a much greater willingness to eat up their greens than we do). Vitamins are mostly pretty simple molecules. Some are carbohydrates, some proteins – they are not related in terms of their chemical structure.

Dietary minerals are how we get hold of the other elements our bodies need, besides carbon, hydrogen and oxygen. For example, our red blood cells contain iron; some enzymes need zinc, copper and selenium; our bones contain magnesium; and phosphorus, sodium and chlorine have multiple uses. Because these elements are necessary for other animals and plants too, we usually get enough of them in our regular diets, but taking supplements is popular too.

Dietary fibre is made of the cellulose that forms the tough cell walls in plants. Even though it mostly passes through our systems undigested, its passage is necessary to keep our digestive systems – especially our colons – in good working order. Some of the ways in which we process foods – turning 'brown' flour to 'white', for example, amounts to removing the cellulose, and in many Western cultures people don't eat enough fibre and have higher rates of bowel trouble.

HARMFUL CHEMICALS

It seems like a paradox, but sometimes the chemicals that can do real damage to our bodies are structurally similar to those that are actually supposed to be there. They disrupt normal body processes, usually by chemical reactions which deactivate, destabilize or otherwise interfere with the body's own molecules. Some substances, which are beneficial and utilized by the body in smaller amounts, can be poisonous in larger quantities.

Poisoning may be a rapid and dramatic process or very gradual and cumulative, and can be severe or mild, depending on the type of poison or toxin and the amount consumed. Poisons may target the nervous system, destroy blood cells, and prevent the liver working, among other unpleasant things. Many poisonous substances can be cleared out by the body over time – alcohol is a good example – but often there is an amount beyond which there's no return – the lethal dose. Some poisons are harmless when ingested, but bodily processes turn them into toxins. Many of the medicines we use are poisons, designed to inhibit a natural process which is happening to excess.

Poisons in the Diet

Some living things manufacture poisons in their bodies, to protect themselves from being eaten. If a bird eats a little fluffy white moth and promptly feels ill because of toxins in the moth's body, it is likely to avoid fluffy white moths in future, so, as a whole, that species of fluffy white moth benefits. We know (and presumably our ancestors learned the hard way) that not every shiny red berry is delicious and nutritious. Such is our accumulated knowledge of what is and isn't safe to eat that people rarely poison themselves this way today.

Pesticides are tailor-made to be poisonous, to kill plants and animals that interfere with the crops we grow, so it's no surprise that they can poison us too. After World War II, the chemical DDT, or dichlorodiphenyltrichloroethane to give it the name that explains why we call it DDT, was widely used as an insecticide. By the 1960s, severe declines in many bird species were linked to its use. The chemical accumulates in animal bodies, meaning that birds that eat poisoned insects carry a DDT 'load' which is then passed on to the next species in the food chain. The effects on birds of prey were particularly devastating. Its use has been banned for decades now but it is an extremely persistent chemical and its residues can still be found in the fat stores of livestock and fish.

Airborne Poisons

We took a look at air pollution earlier. These poisons get into our systems via our lungs. The solid particles don't get through into our bloodstreams but stay in our lungs and slowly clog them up. The gaseous molecules do get into our blood and from there can cause havoc. One of the most dangerous is the odourless and colourless carbon monoxide. At least with sulphur dioxide you know it's there by its horrible smell, and nitrogen dioxide is similarly smelly and has a red-brown colour. Carbon monoxide poisons by reacting with the haemoglobin in your red blood cells, making them unable to carry oxygen.

VENOM

Want to annoy a biologist? Start talking about 'poisonous snakes'. In biology, poison that an animal administers to another via fangs or a sting is called venom and the animal that does it is venomous.

BIOLOGY

HUMAN (AND OTHER) BODIES

CIRCULATION

A human body is rather like a city. The organs are like buildings, some used for storage, some for production, some for processing. The 'workers' and the 'products' – blood cells, hormones, molecules of food and so on – are like vehicles, travelling from organ to organ as required via our circulatory systems – our internal transport networks. The most important of these networks is the one that carries our blood – the cardiovascular system.

OXYGEN AND BLOOD VESSELS

The oxygen we breathe in is needed by most of the cells in our bodies to carry out essential functions. One of the most important 'jobs' of blood is taking this oxygen to where it's needed, then bringing the blood back to the lungs to load up again with fresh oxygen. The main blood

BLOOD COMPOSITION

There are three main constituents of our blood. The red blood cells or erythrocytes are small cells with a squashed disc shape. They contain haemoglobin, which combines with oxygen, and, unsurprisingly, they give blood its red colour. White blood cells or lymphocytes are part of our immune system and their role is to deal with any bacteria that enter our blood stream – they also include platelets, which are involved with blood clotting when a blood vessel is broken. These are all borne in plasma, a pale yellow liquid that's mostly water and is there to carry the cells around – hormones and food molecules also travel 'loose' in the plasma.

vessels carrying oxygenated blood away from the heart to the body are the arteries, which divide up into smaller vessels called arterioles. These divide again into tiny vessels – capillaries. Oxygen exchange happens through the walls of the capillaries – unlike the larger blood vessels their walls are just one cell thick and through them oxygen and other substances can be exchanged.

The de-oxygenated blood proceeds from capillaries to venules (equivalent to arterioles), which in turn join the large veins (equivalent to arteries) that eventually return the blood to the heart.

Arteries are thicker than veins and their walls contract to help keep the blood moving, producing the pulses we can feel in our necks, wrists and other places. Veins are less muscular, but the squeezing of our surrounding muscles help keep blood moving through them (that's why brides and grooms faint at the altar if they stand still too long), and valves prevent blood 'back-flow'.

The Heart

The heart is, basically, a blood pump. As organs go it's rather simple. It is divided vertically into two halves, each with top and bottom chambers. The left side of the heart receives freshly oxygenated blood from the lungs into the top chamber – the left atrium, via the pulmonary vein. The blood proceeds into the lower chamber – the left ventricle, which pumps powerfully to send that blood off to the rest of the body via our largest artery – the aorta. The right atrium receives the de-oxygenated blood via our largest vein – the vena cava. It then travels to the right ventricle, which pumps it off to the lungs via the pulmonary artery to collect more oxygen.

ARTERIES AND OXYGEN

Arteries go out from the heart and veins return to it, so in nearly all cases you'll find bright red oxygen-rich blood in arteries and darker oxygen-depleted blood in veins. The exception is with the pulmonary artery and vein, which travel to and from the lungs respectively.

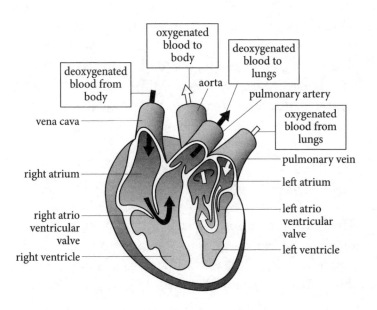

The heart's walls are made of cardiac muscle, which works constantly and unconsciously (good thing too). Ventricles are thicker-walled than the atria, as they do more serious pumping. Valves between atrium and ventricle and ventricle and aorta/pulmonary artery regulate the blood flow. It is the closing of these valves that produces our heartbeat.

OTHER CIRCULATORY SYSTEMS

Alongside our cardiovascular system is the lymphatic system, which also comprises a network of tubular vessels (though it lacks a central pump). It carries lymph – a clear fluid similar to blood plasma. The lymphatic system's role is as a go-between for blood and body cells, moving around such things as newly formed white blood cells, food molecules and excess interstitial fluid (the fluid that surrounds all of our body cells).

CIRCULATION IN OTHER ORGANISMS

The heart of an insect doesn't carry oxygen, only food, therefore its blood is green. The heart itself is a simple muscular tube. Other vertebrates have simpler hearts than birds or mammals, with fish having just a single atrium and ventricle. Plants transport water around their various parts via a network of xylem vessels, while phloem vessels transport food in the form of sap.

SKELETAL STRUCTURE

Bones have four main functions: they are anchor points for the muscles we use to move around, thus providing support and structure to our otherwise rather squishy bodies; they protect our particularly squishy and important organs from damage; they store minerals; and bone marrow is the factory that manufactures our blood cells.

THE HUMAN SKELETON

As we all know, the head bone is connected to the neck bone, and so on. Except that it's rather more complex than that, but the song would need to mention 206 individual bones (in adults – babies have more but some fuse as they grow) to be strictly anatomically accurate. We'll aim for the middle ground and describe the main groups of bones.

The skull protects the brain. The lower jaw or

mandible is a separate part. The tiny ossicles are three bones in each inner ear that help us to hear.

The vertebral column or spine is comprised of thirty-three vertebrae, chunky little bones with holes in their centres through which the spinal cord runs. The column is traditionally divided into four sections – cervical, thoracic, lumbar and pelvic (neck, upper back, lower back and bottom).

Limb bones comprise one long bone in the upper limb (humerus in the arm, femur in the leg), two long bones in the lower limb (ulna and radius in the arm, tibia and fibula in the leg with the kneecap or patella between them and the femur), carpals and metacarpals in the wrist and hand, tarsals and metatarsals in the ankle and foot, and phalanges in the fingers and toes.

Finally, for want of a better name, 'the bones in the middle' – the ribcage protects the main organs, with the sternum or breastbone uniting the upper ribs at the front of the body. The scapulars (shoulder blades) and clavicles (collarbones) connect the arms to the upper body, and the pelvis connects the legs to the lower body while also helping to protect lower-body organs.

HANGING IT ALL TOGETHER

Bones connect to each other via joints, which come in different kinds and allow different degrees of movement. Synovial joints, like the ball-and-socket joint of the shoulder or the hinge joint of the knee, allow very free movement. These complex joints are protected by pads of cartilage and lubricated by synovial fluid. Other bones are joined by bands of cartilage or fibrous tissue, and allow minimal movement.

You'll find essentially the same bone groups in most other vertebrates, though they vary a lot in structure and relative size. For example, the inner ear bones in mammals are found in the jaws of reptiles, where they have a completely different function. Invertebrates have rigid outer coverings instead of internal skeletons (exoskeletons) or do without 'hard parts' in their bodies altogether.

BONE BUILDING

Crack a bone in half and you'll see that the outer layers are hard and mineralized, but inside it's soft and spongy.

The outer layers are made of inactive bone-forming cells (osteoblasts). These layers are dense, compacted and contain lots of calcium and phosphorus. The spongy bone inside contains active bone-forming cells, which mainly make the tough protein known as collagen. It has a generous blood supply and includes the red, soft 'bone marrow' that manufactures blood cells of all kinds.

MUSCLES AND SKIN

If you want to turn heads on a nudist beach, you'll want your muscles and skin to look their best. Appearances aside, the two form distinct systems. Your muscles move you around and your skin is your interface with the world outside – it provides a physical barrier, but also enables you to take in some of the things your body needs and get rid of some of the things it doesn't.

How to Move

Two of the three kinds of muscle – cardiac muscle in the heart and smooth muscle in places like our digestive system – are involuntary and carry on doing their thing as long as we're alive. Skeletal muscle, however, is voluntary muscle – we (usually) decide when we want it to do something, albeit often on a subconscious level.

When a muscle is contracted, it shortens. In the case of the biceps muscle, which is on the top of the upper arm and is connected at the shoulder and the top of the forearm, its contraction causes your arm to bend. Conversely, the triceps, which is on the back of the upper arm, causes your arm to straighten when it is contracted. The abdominal muscles in your tummy cause your middle to curl forwards when contracted – muscles in your lower back straighten you out again. Most body movements involve several muscles at the same time, working to a greater or lesser degree (just as those diagrams on the weights machines at the gym show you). Some other muscles provide stability rather than movement when contracted.

The human body contains more than 600 skeletal muscles. All of them are made of bundles of muscle fibres, enclosed in a tough membrane sheath. At their ends they become tendons, which attach to bones. The contraction within the muscle fibres is caused by bonds forming and breaking between two protein types – actin and myosin, in response to electric signals from motor nerves – the ones that send messages from the brain to the muscles when we want to move.

SKIN COLOUR

The amount of melanin in our skin gives us our skin colour. Darker skin is commoner in hot places, as melanin acts as a sunscreen, protecting against dangerous ultraviolet radiation. However, for the same reason high levels of melanin make us less able to make vitamin D (which is manufactured in the skin in response to sunlight), so in less sunny places people tend to have less melanin. Dark-skinned people who live in northern places with less sunshine need to make sure they get sufficient vitamin D in their diets, just as pale-skinned people need to take extra care to protect their skin with sunscreen in hot countries.

OUTSIDE EDGE

The largest human organ is the skin. It is a flexible, continuous covering over nearly all of the body, keeping bacteria out, regulating our temperature and water content, helping us make vitamin D and providing sensory information about the world around us.

The outer layer of skin – the epidermis – is composed

of dead epithelial cells, which are continuously worn away and replaced. Hairs grow out of it, and pores within it are outlets for sweat glands. At the bottom of the epidermis is the pigment layer where the skin pigment melanin is found. Further down is the dermis layer, wherein are the follicles of the hairs with their associated erector muscles and sebaceous glands, the sweat glands themselves, blood vessels and sensory nerve endings of various kinds. Under these is a layer of fat – the subcutis – which provides insulation.

Sebaceous glands produce an oily substance called sebum, which helps keep skin and hair soft and flexible.

FUR COATS

Each of a mammal's body hairs has a little muscle which can make the hair stand on end. While this is pretty useless in humans, in furry critters it can make a huge difference to retaining and losing heat. Fur standing on end also makes the critter look larger, which can help it scare off a rival or potential predator. Fur is also a handy 'canvas' for colours, patterns and markings.

Sweat glands release salty water, which helps to cool us down by evaporation as well as regulating our salt and water levels. The blood vessels close to our skin can widen when we need to lose some heat – that's why we get flushed after exercise – and they can constrict to retain extra heat when we're cold. Sensory receptors in our skin detect pressure, vibration, heat, cold and pain. We can absorb some substances through our skin (sadly not always things we would want to).

NERVOUS SYSTEM

Boo! When startling things like that happen, you become very aware of your nervous system. But it's also busily working away in the background all the time, because it controls a vast array of conscious and unconscious bodily processes – in fact, all of them.

A SINGULAR CELL

The nerve cell or neuron is, shape-wise, quite a departure from the generic blob-shaped cell with a central nucleus. At one end is the cell body with a nucleus – it has numerous long branched projections called dendrites. Another very long projection forms the axon, down which the electrical charge of a nerve impulse travels. The axon is insulated with wads of fatty myelin, enabling the nerve impulse to travel faster as it jumps from gap to gap between the wads. The other end of the cell is formed by the 'foot' of the axon, where there are several bulb-tipped terminal branches. These release a chemical (neurotransmitter) that activates receptors on the next cell, so the impulse continues to its destination. Nerves are bunches of axons gathered together like cables.

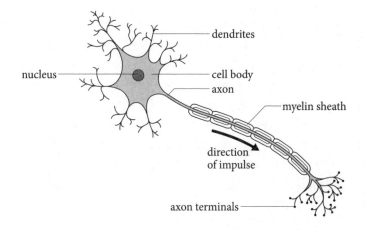

Brain Box

The brain is the centre of the nervous system. It is basically a big mass of neurons, up to thirty-three billion of them, along with a larger quantity of glial cells that have various roles but basically take care of the neurons. All of this is organized into structures and regions with distinctly different functions. The wrinkled, convoluted domes of the cerebral cortex include the parts concerned with perception (interpreting sensory information) and conscious thought, and are proportionately much larger in humans than other animals. The more tightly wrinkled round bit at the back under the cerebral cortex is the cerebellum, which is mainly concerned with automatic physical movement (things most of us can do without thinking about it, like standing up, walking and running). Other brain structures nestle deep inside and deal with a wide variety of other unconscious functions.

The brain stem emerges from the base of the brain and becomes the spinal cord, a thick bundle of nerves that travels down the centre of the vertebral column, sending off branches to the various body parts as it goes. Brain + spinal cord = the central nervous system. The outbranchings of the spinal cord form the peripheral nervous system.

SOMATIC AND AUTONOMIC

The peripheral nervous system has two components. There is the somatic nervous system, which deals with voluntary body movements and sensory inputs, and there is the autonomic nervous system, which handles mostly unconscious stuff like heartbeat, digestive processes and so on. The two systems use different nerve pathways.

FIGHT OR FLIGHT?

Back to that startling 'Boo!'. What happens to your body when you get a fright? The autonomic nervous system kicks in – specifically, the part of it known as the sympathetic nervous system. It brings about a whole range of physiological changes designed to get you ready to deal with the danger or to run away very fast. Adrenaline is released, to get your heart rate up. Your pupils widen to better see what's happening, while your digestive tract slows down, so blood is freed up for the muscles, where it's needed. The counterpart system, the parasympathetic nervous system, basically does the opposite and calms everything down again.

DIGESTIVE SYSTEM

Our bodies are made up mainly of proteins, fats, carbohydrates and water. We are constantly using up and breaking down these substances in physical processes, such as growth, replacing broken bits and moving around, so we need to keep on putting more in, and we do this by eating plants and other animals – our digestive tracts do the rest.

MAKING TRACTS

The journey from one end of the human digestive tract to the other is a circuitous one, covering about 6.5 metres in the average adult man. The journey time is, if anything, even more surprising, normally taking thirty-five hours or more. All that time and distance is necessary to convert your fish and chip supper into a collection of molecules small enough to be absorbed into the bloodstream.

Swallowed food goes down your oesophagus to your stomach, which vigorously churns it around and steeps it in strong acids, and after three to five hours of processing there it has all moved on to the small intestine. This takes about three hours, then it's on to the colon where transit takes up to forty hours (but can be a lot less). The pancreas, kidneys, liver and gall bladder all play a role in the process but are not part of the tract.

ENZYMES

Along the digestive tract, enzymes are released to
help break down the food. Enzymes are proteins,
which break down food into successively smaller
molecules. Different enzymes work on different
nutrients – proteases break proteins down to amino
acids, amylases turn complex carbohydrates into
simple sugars and lipases break up fats into fatty
acids. The first enzymes are produced by the tongue,
and then the pancreas releases more into the food
as it leaves the stomach. Acids produced in the
stomach and, later, more that are formed in the liver
and released by the gall bladder help the enzymes to
access their target food molecules.

Soaking it up
The main role of the small intestine is food absorption.
The inner walls of this long and fantastically twisty tube
are lined with finger-like projections called villi, through
which food molecules can pass straight into blood
capillaries. Most of the breakdown of food by pancreatic
enzymes takes place in the twenty-five centimetre-long

duodenum, the first short stretch of the small intestine, while the two-metre-long jejunum does most of the nutrient absorption and the 3.5 metre ileum the rest.

Waste disposal

By the time the food reaches the colon, practically all of the useable nutrients are gone. The colon's main role is to absorb excess water from what's left, though it also takes in a few vitamins that the small intestine doesn't. Bacteria that live in the colon break down dietary fibre here, and quite a lot of them end up becoming part of the final waste product themselves. When it comes to urinating, excess water in the blood is filtered out by the kidneys, and leaves the body via the urethra after a decent quantity of it has built up in the bladder.

DID YOU KNOW?

Birds may be toothless but to compensate they have a couple of extra organs in their digestive tracts. The crop, an enlarged, stomach-like section of the oesophagus, stores food prior to digestion – its contents may be regurgitated up to feed the baby birds. The gizzard is a muscular grinding 'second stomach' – your budgie eats grit to provide its gizzard with extra seed-crushing power. The stomachs of cows and other ruminant animals are divided into four chambers – food swallowed into the first (the rumen) may be brought back up again for some re-chewing before being swallowed again and moved on to the other chambers for digestion.

REPRODUCTIVE SYSTEM

Here's the bit you probably turned to first in your school biology textbook, to check for naughty illustrations. Like most animals, we are very highly motivated to have sex – passing on our genes is as much a part of the game of survival as is finding enough to eat, drink and breathe, though usually considerably more difficult to accomplish.

Boys' Toys

The male reproductive system is the simpler of the two so let's begin there. There are two testes* which hang outside the body in the sack-like scrotum – their function is

---

*This is the correct plural of 'testicle', though only the worst pedants will pretend not to know what you mean if you use 'testicles' instead.

the manufacture of sperm cells and they require lower temperatures than are found within the body to best accomplish this. A narrow tube, called the vas deferens, carries sperm to the penis when a man ejaculates. On the way, the prostate gland secretes various fluids, which are added to the sperm to produce semen.

WOMANLY PARTS

Most of the female reproductive apparatus is internal. Of the external bits, the clitoris is a small and highly sensitive organ, analogous to the penis but with the sole function of making sex fun. The ovaries, which are analogous to the testicles, are located inside the lower abdomen. They contain the eggs – female sex cells. Females don't constantly make new eggs but at birth already have all that they'll ever produce. Once a month, a single (usually) egg matures in one of the ovaries and is released into the Fallopian tube. The two tubes connect to the top of the uterus or womb, which becomes lined with thick, blood-rich spongy material in the days leading up to the arrival of the egg. If the egg isn't fertilized, it is released through the vagina, along with the uterine lining – this is menstruation. If the egg has been fertilized, it implants in the uterine lining and begins to grow.

Splitting cells

Normal cell division is called mitosis. It begins with the duplication of all the chromosomes (sets of genetic material) in the cell nucleus, and results in the formation of two identical cells from the single parent cell. This process happens constantly in many different parts of our bodies. Things are different with sex cells. Their division is called meiosis, and the parent cell splits into four daughter cells, each with half the chromosomes of the parent. When two sex cells unite in fertilization, they produce a single cell with a full set of chromosomes, half derived from the mother and half from the father.

BIRTH AND BEYOND

When a baby is ready to be born, the amniotic sac breaks, unloading all that fluid, and leaving everyone in no doubt as to what's to come. The hormone oxytocin stimulates the uterus to contract, and after (sometimes many) hours of increasing contraction, the opening in the neck of the uterus (the cervix) is wide enough to let the baby through. Complications with this process may necessitate a caesarian section, where the baby is removed via an incision in the mother's belly. The mother's breasts soon start to produce milk – a fatty nutritious substance which is all the baby requires in the way of food for its first months of life.

MAKING BABIES

When the man ejaculates during sexual intercourse, the sperm travels into the woman's uterus. Sperm cells, with their long tadpole-like tails, are reasonably competent swimmers as cells go, though can't survive too long in the hostile acidic environment they find themselves in. If the timing is right and there is an egg travelling down one of the Fallopian tubes, the sperms can detect it by temperature. They surround the egg and one lucky individual penetrates its outer layers, resulting in the fusion of the chromosomes of both cells and the beginning of a new human life. By the time it implants, the new cell has already divided many times. Over the next nine months the cell division and specialization continues, and it develops into an embryo, foetus and eventually a baby. A placenta also develops – an organ that allows the blood of the growing foetus to exchange nutrients and other essential substances with the blood of the mother, and an amniotic sac – a membrane containing cushioning amniotic fluid – forms around the foetus.

RESPIRATORY SYSTEM

Who doesn't enjoy breathing? Our cells need oxygen to release energy – just as importantly, we need to get rid of the carbon dioxide that's a waste product of that same cellular process. The air in Earth's atmosphere is mostly nitrogen, which we breathe in and out without anything happening to it, but there's 21% oxygen in it too, which is sufficient for humans (too much oxygen is actually toxic to us).

NOSE, THROAT AND LUNGS

Air enters our lungs via the nose and/or mouth, though the nose generally does a better job. It is lined with tiny hairs that trap some of the larger airborne particles that you wouldn't want to inhale, and the mouth doesn't. The airways from nose and mouth join to enter the trachea or windpipe, a tough tube which runs down the throat into the chest. Here it divides into two main bronchi, which enter the two lungs.

The lungs are big squashy bags. We've all seen distressing images of lungs blackened from decades of smoking – healthy lungs are a more appealing light pinkish-grey. The right lung is shorter and wider, and divided into three lobes, while the left has only two – this difference is because of the way the liver and heart

137

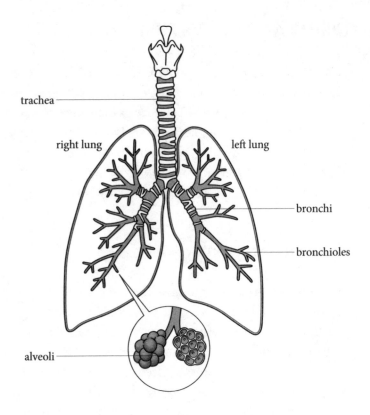

trachea

right lung   left lung

bronchi

bronchioles

alveoli

fit around the lungs. The bronchi continue to branch off into smaller and smaller tubes (bronchioles) inside the lungs, which eventually meet the alveoli – the structures where the gas exchange actually happens.

Gas Bags

Alveoli look like little clusters of tightly-packed grapes. There are 150 million of them in each lung. Gas exchange takes place on their entire surface area, which amounts to a huge seventy-five square metres if you took the trouble to flatten them out. They have a very rich blood supply in the form of numerous tiny capillaries. Here, inhaled oxygen is transferred to the red blood cells, while waste carbon dioxide is released from the blood to the alveoli to be exhaled.

WHOSE AIR IS IT ANYWAY?

Exhaled air normally has about 4.5% less oxygen in it than it did just before you inhaled it, and it contains about 4.5% carbon dioxide, versus 0.03% carbon dioxide in inhaled air. Each breath we take draws in between four to six litres of air, and going by the number of humans on Earth and the amount of available air, we can deduce that every breath we take contains at least some molecules that have previously been inhaled by Socrates, Caesar, Beyoncé or whoever you care to mention.

SENSORY SYSTEMS

Our senses are the means by which we get information about our environment. The five 'traditional' senses are sight, hearing, smell, taste and touch, but most biologists agree we have a few more as well, like thermoception (sensing changes in temperature) and equilibrioception (sense of balance). The outside information is detected by receptor cells in the sensory organ (eye, skin, ear and so on) and these are all wired into our nervous systems, so the information gathered is translated into impulses along nerves all the way to our brains, where we interpret the data and make decisions about what to do with it.

Some other animals rejoice in extra senses, such as electroception (the ability to detect electric fields, found in various fish, and duck-billed platypuses) and magnetoception or magnetoreception (the ability to detect magnetic fields – found in migratory birds).

EYES

The simplest kind of animal eye can distinguish light from darkness – it switches from 'off' to 'on' when a certain minimal level of light is present, but has no ability to detect graduations in light levels. Not much use, you might think, but if you're swimming about in the sea and it all

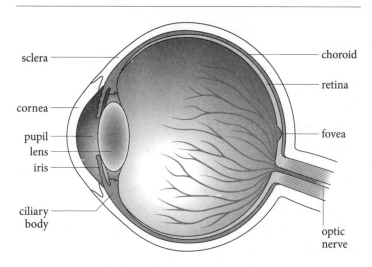

sclera

cornea

pupil
lens
iris

ciliary
body

choroid

retina

fovea

optic
nerve

suddenly goes dark, that probably means something big and dangerous is approaching and you should leave the area as quickly as possible.

Human eyes are much more sophisticated than that. We can see a wide range of colours and resolve lots of detail from the picture of visible light that hits our retinas. We can also focus at close range and long-distance, can cope quite well in low light and can quickly detect movement.

From the front, on a non-dissected human, the bits of the eye you can see are a bit of the sclera (the 'white' of the eye; a tough covering – it becomes transparent at the front where it's called the cornea), the iris (the coloured part; a round sphincter muscle which contracts or relaxes to change the size of the pupil) and the pupil (the black bit; actually just the hole at the centre of the iris, through which light passes).

Once through the pupil, the light radiation hits the lens, a transparent structure which changes shape to focus the light before it reaches the retina – the inside layer of the eyeball which contains the special light-sensitive nerve cells. These cells contain chemicals that react if certain wavelengths of light reach them. They come in three kinds – rods, cones and photosensitive ganglion cells.

Rods are very sensitive to light of all colours, so help us detect movement quickly – especially at the edges of our vision.

Cones are sensitive to one of three different colours of light – red, green or blue. Between them the cones supply our colour vision, and they are concentrated at the fovea – the part of the retina that aligns with whatever we are looking at directly.

Photosensitive ganglion cells (only discovered in the 1990s – if you haven't heard of them that's why) detect slow, long-term light changes, so help provide length-of-day information.

Chemical reactions in the retinal cells act as triggers to adjacent nerves, causing them to 'fire', and all these impulses are channelled down the optic nerve which exits out of the back of the eye. The point where it leaves gives us an optical blind spot – a point on the retina where there are no light-sensitive cells. We are very good at unconsciously compensating for this, however, so that

FIND YOUR BLIND SPOT

Here is one of several optical illusions that will show you that you do indeed have a blind spot.

Look at the picture from a distance. Cover your right eye and stare at the right-hand circle (containing the X). Now move your head very slowly towards the image. At some point, the left-hand circle (containing the spot) will disappear, as its position coincides with your blind spot. You'll see the black and white square pattern instead – that's your brain filling in the gaps by 'guessing' what might be there in the part it can't 'see'.

we hardly ever notice it.

The optic nerves from each eye have a long journey to the occipital lobes of the brain, the parts that deal with sensory information – they're right at the back. Halfway along, the nerve fibres originating from the inner section of each retina cross over and end up in the opposite side of the brain. This crossover means that each brain hemisphere deals with half of the information from both eyes.

EARS AND SKIN

Sound waves enter our ears and cause the eardrum, a thin membrane, to vibrate. These vibrations pass through the ossicles, which are three tiny bones – the malleus, incus and stapes (aka the hammer, anvil and stirrup) – they increase the pressure of the sound waves. From there the waves reach the liquid-filled cochlea in the inner ear, a coiled-up tube structure which contains the receptor cells that convert physical vibrations into electric signals that travel to the brain for interpretation.

The hearing receptor cells in our ears are hair cells, which have fine projections called cilia that detect vibrations in the fluid inside the cochlea. Although different in structure, some of the receptor cells in our skin also detect vibration or pressure, and it's often said that snakes and some other animals 'hear through their skin'.

SMELL AND TASTE

The receptor cells in our noses and the taste buds on our tongues are chemoreceptors – they are activated in the presence of certain chemical molecules. For a long time the human tongue was thought to be sensitive to just four tastes – sweet, salty, sour and bitter – with all the other nuances of flavour actually provided by the nose, which is sensitive to a much wider array of smells (though not nearly so many as some other mammals such as dogs). A fifth taste sensation was identified in the early twentieth century – umami or 'savouriness'.

SENSORY MULTI-TASKING

The inner ears also contain our vestibular system, which helps us keep our balance. It consists of three fluid-filled looping tubes – semicircular canals. The movement of the fluid inside them as our heads move around tells us which way up we are and whether we're about to fall over.

CELL BIOLOGY

STRUCTURE OF A CELL

You're made of cells. So is your cat, and so are your eyelash mites, and the mushrooms in your omelette – you get the idea. Life-forms like us that are made of one or more complex cells make up one of the three primary divisions or domains of life on Earth – the Eukarya ('good nucleus'). The other two are both kinds of bacteria, which have a much simpler structure.

A GENERIC ANIMAL CELL

In our bodies, cells are specialized to do different jobs. Not all are spherical-ish blobs. Nerve cells are long and thin, red blood cells are like doughnuts with a squashed bit rather than a hole in the middle, sperm cells look like tadpoles and so on. However, most cells have the same general working parts or organelles, as shown in this (rather simplified) diagram.

Now, what do all those bits do?

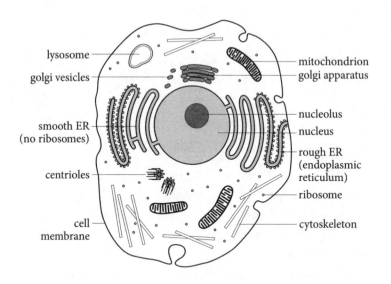

The cell membrane might seem the most trivial part, but it's a lot more than an organic bag to keep everything together. It is semi-permeable, which means it lets some molecules and chemical ions in and out – but not all, and not always at the same time – some of this transport is passive and happens by diffusion, but there is also active transport via protein pumps and ion channels.

The cytoplasm, or protoplasm, comprises the organelles of the cell plus the fluid (cytosol) in which they are suspended. The cytosol is mostly water with dissolved salts and a few other molecules – it also contains a network of protein fibres, the cytoskeleton, which supports the shape of the cell.

Mitochondria (singular – mitochondrion) are rod- or spindle-shaped organelles that make the adenosine triphosphate, which supplies the cell's energy needs. They are membrane-bound and contain a huge number of different types of protein. The theory of endosymbiosis says that mitochondria were originally free-living bacteria, but became engulfed by other cells.

Ribosomes are small round organelles that manufacture proteins from amino acids. They contain ribosomal RNA, which provides the code needed to put the amino acids together in the right order to make a specific protein.

The endoplasmic reticulum is a network of membrane-bound tubes and vessels inside which various functions take place, including synthesizing fats, neutralizing toxins and 'receiving' newly formed proteins from ribosomes.

The nucleus is the largest organelle in the cell. It is bound in a nuclear membrane, through which some small molecules can pass. The cell's DNA lives within the nucleus, and is duplicated here when the cell divides. The nucleolus is where ribosomal RNA is made, which is the main constituent of ribosomes.

The Golgi apparatus deals with collecting newly synthesized molecules from the organelles that make them, and secreting them out of the cell.

The barrel-shaped centrioles have an important role in cell division.

Lysosomes contain enzymes to digest the remains of 'dead' cell components, and any virus particles, bacteria or other relatively large chunks of organic matter that the cell has engulfed.

The plant cell

Cells in plants have a lot in common with animal cells, but there are, of course, many differences too. One of the most obvious differences is the presence of a tough cell wall, which surrounds the cell membrane and often gives the cell a more rigid, angular shape. Another is the presence of chloroplasts, small green organelles which carry out

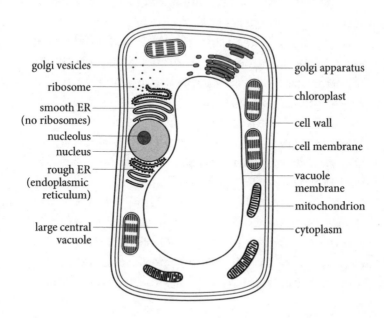

golgi vesicles — golgi apparatus

ribosome — chloroplast

smooth ER (no ribosomes) — cell wall

nucleolus — cell membrane

nucleus

rough ER (endoplasmic reticulum) — vacuole membrane

— mitochondrion

large central vacuole — cytoplasm

the process of photosynthesis. A third difference is that plant cells often contain very large vacuoles – membrane-bound, water-filled compartments that help keep the cell shape rigid as well as dealing with waste products. Animal cells may contain small vacuoles, but in plants they can take up most of the space inside the cell.

PHOTOSYNTHESIS

Solar panels? Plants invented them first, or at least came up with a way to harness the light energy of the sun to create their own energy – photosynthesis. The light of the sun is used to power a chemical reaction between water and carbon dioxide, which makes a useable energy-releasing food-stuff for the plant and releases oxygen as a by-product. Blue-green algae, or cyanobacteria, are the simplest organisms to carry out this process. They were among the first life-forms to appear on Earth, with signs of their existence dating back 2.8 billion years, and were probably responsible for establishing the oxygen-rich atmosphere we know and love today.

As we saw in the previous section, modern plants contain organelles called chloroplasts in their cells. They look suspiciously like cyanobacteria, and support the theory of endosymbiosis – that other cells engulfed the cyanobacteria, where they continued to photosynthesize and to supply food that the engulfing cell was able to use, thus forming the first complex eukaryote cells.

Today, modern plants, algae and cyanobacteria collectively trap a jaw-dropping hundred terawatts of energy from the sun (or a hundred terajoules a second), which is more than six times as much as is consumed by the entirety of human civilization. They also use up a hundred billion metric tonnes of carbon from atmospheric carbon dioxide. Without them, we would be comprehensively doomed.

How it Works

The chemical equation for photosynthesis looks like this:

$$6 \ CO_2 + 6 \ H_2O \ (+ \ energy) \longrightarrow C_6H_{12}O_6 + 6 \ O_2$$

The energy comes in the form of photons – light energy from the sun, which is captured by the chlorophyll pigment inside the plant's chloroplasts. The carbohydrate that's produced is glucose, the simplest form of sugar, and the plant stores that in its tissues, while the oxygen is released back to the atmosphere.

ADP, ATP AND ENERGY

The photosynthesis reaction uses up energy and produces a more complex compound from the reactants. Reactions of this kind are called anabolic. The reverse of this, where a compound is broken down into simpler products and energy is released, is called catabolic. The catabolic reaction of aerobic respiration, which happens in plants and animals alike, is the way living cells use oxygen and glucose to release energy to use as required*. The reaction as an equation looks like this:

$$C_6H_{12}O_6 + 6\ O_2 \longrightarrow 6\ CO_2 + 6\ H_2O\ (+\ energy)$$

If you've been paying attention, this should look familiar. It's the reverse of the photosynthesis reaction. The key difference is the nature of the energy. Energy input in photosynthesis is light energy, the energy output from respiration is chemical, and is carried by the molecule ATP (adenosine triphosphate).

ATP consists of the nucleotide adenosine (which also happens to be a component of DNA) with three phosphate groups (monomers of oxygen and phosphorus) attached. When respiration occurs, molecules of the stable ADP

---

*In biology, this is what respiration means. Non-biologists often incorrectly use it solely to mean breathing, which is very naughty of them.

GIVE AND TAKE (AND TAKE SOME MORE)

We often imagine an oversimplified picture of gas exchange in the natural world, with plants using up carbon dioxide and releasing oxygen, and animals using the oxygen and releasing carbon dioxide, and on and on forever in happy equilibrium. However, as we saw above, plants use up oxygen too in the same respiration reaction as animals. So, the main difference between animals and plants is actually that plants synthesize their own food supply, which happens to use up atmospheric carbon dioxide.

(yep, adenosine diphosphate) which have two phosphate groups, are turned into the unstable ATP by the addition of a third phosphate group, via an unstable, high-energy chemical bond. When energy is required, it is released by the breaking of that bond, which causes the breakdown of ATP back into ADP. Sorry about all the chemistry.

HORMONES

When people talk about feeling 'hormonal', they usually mean specifically the hormones that control the female menstrual cycle. But you could just as well describe the feeling that would result if you opened your front door to find an axe-wielding crazy person standing there as 'hormonal'. Ditto the feeling of being very relaxed and about to nod off to sleep. Hormones regulate a whole range of bodily processes, many of which happen to be associated with particular feelings as well as with essential things like our sleeping-waking cycle, growth, reproductive mechanisms and our physiological response to danger.

Hormones are secreted into the blood stream, where they travel to other parts of the body and cause cells to alter their metabolism (their chemical processes) in some way. That includes the rate at which they produce hormones – in other words, hormones regulate the production of other hormones.

Types of Hormones

All hormones are molecules, but they vary widely in shape, size and composition. All are either protein-based (amine or peptide hormones) or derived from fats (steroid hormones). Hormones are released by endocrine glands, of which there are several in the body. Some organs are nothing but endocrine glands, such as the adrenal gland, one of which sits on top of each kidney. Other organs, such as the pancreas, count their endocrine functions among a wider array of talents.

Let's look first at one hormone that everyone's heard of – testosterone. It is secreted by the testes in men and the ovaries in women, topped up by small contributions from the adrenal gland. Men produce on average fifty times more of it than women, and it is the key hormone in establishing and maintaining 'maleness' in a human body, from conception to maturity. Embryos, as you probably know, all begin with a female-type body plan, but male embryos start to produce testosterone which turns their genitals masculine. In puberty, testosterone causes changes like armpit hair growth and, alas, spots in both sexes, and extra hairiness, voice deepening and shoulder-broadening in boys. In adults (men and women) it helps regulate and maintain the libido.

So, that's a hormone that has wide-ranging effects across several organs, varying according to your stage of life. Another you'll also have heard of is insulin, which is secreted by the pancreas. This has a much more specific effect, causing cells in our muscles, fat and liver to absorb glucose from our bloodstream – its release is triggered when our blood sugar rises after eating. Without enough of it, the glucose accumulates in the blood, causing a range of symptoms, which we call diabetes.

How Hormones Work

Different classes of hormones work in different ways. Peptide hormones form a particular folded shape, which fits into a receptor site on the cell membrane of their target cells, thus activating new metabolic processes inside the cell. Steroid hormones actually enter the target cell and bind to receptors in the cytoplasm. The receptors carry the hormone into the cell nucleus, where the hormone activates the genes to make the particular protein that's required to complete the 'job' of the hormone.

EVOLUTION AND ENVIRONMENT ECOLOGY

THE ORIGINS OF LIFE

Until we invent a reliable forwards-and-backwards time machine, we can't be sure how life first began on Earth – the same problem that we have when contemplating the beginning of the universe (if it had one). We can theorize possible scenarios and even attempt to recreate them in the lab, but we may never have the same degree of confidence about our theories as we do about what is observable in the here and now. For that reason, many people prefer a supernatural explanation for problems like this, but this is a science book so we're going to stick with looking at naturalistic ideas.

ABIOGENESIS

Based on what's emitted from volcanoes today, we can be reasonably confident that the pre-life atmosphere of Earth would have contained a mixture of gases, mainly water vapour but also including methane, ammonia and hydrogen. Apply energy to this mixture with a lightning strike and reactions could occur between the gases, producing simple organic molecules.

The scientists Stanley Miller and Harold Urey tested this idea by applying electric sparks to a mixture of the aforementioned gases and seeing what happened. The result was the generation of quite a few different organic molecules, including sugars, fats and amino acids. All the building blocks necessary to form nucleotides – the basis of the replicating molecules DNA and RNA – were formed. Other subsequent experiments, have yielded similar results, and some have produced whole nucleotides.

RNA World

Our cells contain RNA (ribonucleic acid), which has various functions, including acting as a messenger to carry the coded information in DNA from the cell's nucleus to the ribosomes, which make proteins. Like DNA it is a chain of nucleotide molecules – four different kinds arranged in various different orders, and so has similar information storage properties and self-replication abilities to DNA. However, it has only one strand rather than two like DNA, so is a smaller molecule, able to go places where DNA cannot.

The theory of 'RNA World', first proposed by Walter Gilbert in 1986, hypothesizes that the first 'life' on earth was a large quantity of free-living RNA molecules of various different configurations, all replicating away. RNA can behave as an enzyme, potentially helping to catalyze (speed up) other chemical reactions, which could provide it with additional chemical building blocks. Those configurations of RNA that were most successful at replicating themselves and at breaking down other chemicals would survive better than the rest.

The theory says that in due course the double helix of DNA would have formed, and survived well because it is inherently much more stable than RNA. From free-living RNA and DNA to the first simple cells is not so much of a stretch – a virus is little more than some DNA or RNA strands held inside a protein coat, and a bacterium a little more complex again. The theory of endosymbiosis (when one simple cell engulfs another and the engulfed cell continues to 'work' for both cells) is the popular explanation for the origin of complex cells*.

---

*Why don't we still have free-living RNA today, then? Simply because all the bigger and more complex things that have evolved since would eat it.

THE EVOLUTION OF THE EUKARYOTIC CELL

We already touched on how the first complex eukaryotic cells – the kind that has a nucleus and other organelles,

DOMAINS OF LIFE

Before we go any further, we need to take a quick look at the primary divisions of life on earth – or at least cellular life (in other words, viruses and other simple 'possible-life' entities are excluded).

The first subdivision of cellular life is into domains, of which there are three – Archaea, Bacteria and Eukarya. Eukarya are the ones made of complex cells – at least one cell but often many, each cell containing a nucleus and lots of other organelles. Archaea and Bacteria are simple single-celled organisms, smaller in size and with no nucleus and very few organelles. Modern bacteria, including cyanobacteria, belong in the domain Bacteria. However, one group of somewhat oddball bacteria were found to be profoundly different and they are now classified in an entirely separate domain – the Archaea (singular – archaeon).

and the kind that you and I are made of – evolved. Time for a closer look at the theory of endosymbiosis. We can see many endosymbionts around today – organisms that can only live inside another organism, with both benefitting from the association. Examples include the bacteria that live inside the roots of pea plants, obtaining nitrogen for their hosts, and friendly bacteria that live inside the intestines of animals, helping them metabolize their food.

ENDOSYMBIOSIS

The theory of endosymbiosis says that an archaeon engulfed some bacteria cells in the first stage of evolution of the eukaryotic cell – the archeon became a eukaryotic cell and the bacteria cells became mitochondria or choloroplasts inside it. This idea is borne out by the fact that DNA in our cells' nucleus is similar to that of Archaea.

So, the mitochondria in a complex plant or animal cell are thought to have descended from free-living bacteria. Likewise, chloroplasts in plant cells probably originated as free-living cyanobacteria – aka blue-green algae. Both chloroplasts and mitochondria contain their own DNA, which is different from that found in the nucleus of the cell they're in. Their membranes are structured like those of bacteria. Both contain their own ribosomes, which are bacteria-like and distinct from the ribosomes at large in the cell itself. They reproduce themselves by dividing, just as bacteria do. It has been suggested that other cell components were also once free-living organisms, but there is not so much evidence for them as there is for mitochondria and chloroplasts.

AND THE REST?

There's a lot more to a eukaryotic cell than its mitochondria and (in plants) its chloroplasts. The nucleus, most important organelle of all, is of uncertain origin, but some biologists believe that the DNA it contains is that which belonged to the 'engulfer' archaeon cell when endosymbiosis occurred, while others think it came from a separate virus which also became engulfed.

Ribosomes are the only organelles which are found inside Archaea and Bacteria cells. Their origin therefore predates that of the eukaryotic cell itself, and indeed the free ribosomes in a eukaryotic cell's cytoplasm are like those of an archaeon, while the ribosomes inside mitochondria and chloroplasts are like those of a bacterium.

Most other cell organelles are generally similar to each other, and are thought to have started out life as inlets of the main cell membrane, gradually becoming more specialized over time. The driving force behind their changes was simply how well they helped the cell to survive and propagate itself – natural selection, which we'll look at in more detail in the next section.

MUTATION AND NATURAL SELECTION

From such simple beginnings came the whole diversity of life on Earth, guided only by the natural processes of genetic mutation and natural selection – in other words, evolution. Like abiogenesis, the theory of evolution is much challenged by those who prefer a supernatural explanation for how we got here, but evolution has much more solid science to support it than abiogenesis does. To argue against it is akin to disputing the theory of gravity, in terms of the amount of evidence that you would need to explain away by some other means.

HOW IT WORKS

Evolution happens to a population of animals when there are two conditions. The first is that the population of organisms has variety – all individuals are slightly different. The second is that there are environmental pressures on the animals – only those that function best in their environment will survive. For a real-world example, consider a population of gazelles. Lions eat the gazelles, so the gazelles that run faster are more likely to escape the lions and live long enough to breed, passing on their fast-running genes to their offspring. So the proportion of speedy runners increases in the population as a whole.

Variety in a population is introduced by mutation. When a single-cell organism reproduces, it splits in two and its DNA is replicated, but small mistakes – mutations – may occur in that replication. If that happens, the new cell is genetically different to the parent cell. In species that reproduce sexually, with each organism's genes coming from two parents, more genetic variety is added. The complete set of genes, or genome, are the instructions for 'building' the organism – a new genome produces an organism that's built differently to its parent or parents.

The word 'mutation' has negative connotations, as most of the mutations we know about are harmful. Mutation is random and undirected – a mutation is as likely, if not more so, to have a damaging or neutral effect as it is to have an advantageous one. That's where the other part of the process – natural selection – comes in. This process ruthlessly weeds out all those individuals that aren't well equipped for survival, so that only advantageous mutations are preserved in a population – the survival of the fittest. We do a similar thing with our domestic

THE PEPPERED MOTH

One of the most famous examples of natural
selection in action concerns a fairly large fluffy moth
called the peppered moth. Most peppered moths are
a pale grey, but a genetic variant with blackish wings
occurs from time to time. In industrialized areas
in the late 1800s and early 1900s, the black form
was found to be much more common than the pale
form.

Peppered moths fly by night, and in the daytime
they rest on tree branches. The dappled grey colour
of the pale form provides good camouflage against
a lichen-covered branch, while the black form
stands out like a red wine stain on a sheepskin rug
and is quickly spotted by predators. However, when
industrial soot coated city trees, the tables were
turned and it was the black form that had the good
camouflage. We've cleaned up our act quite well
since then and the black form is rare again.

animals and cultivated plants, breeding only from those
that show the traits we want to keep and develop. This

process – artificial selection – is responsible for the huge variety of dogs, chickens and chrysanthemums that live on Earth today.

DIVIDE AND CONQUER

Where one species occupies a large, environmentally diverse area, there can be different selective pressures on the different parts of the population. Given consistently varied selective pressures over a sufficiently long time, a population can split into two distinct types. One of the most widespread species of birds in the world is the peregrine falcon. Peregrines live on all continents except Antarctica, and in all kinds of environments except the Poles, the highest mountains and the deepest rainforests. However, not all of them are the same. Peregrines from the Arctic are larger and paler than those in the tropics. Being pale is good for snowy environments, and being large is better than being small when you need to retain heat in cold climates, therefore the larger and paler birds survived best in the north and came to dominate the peregrine populations there.

The next stage along is where a population becomes physically divided, with no interchange between the two. The mosquito *Culex pipiens* is common in the UK. At some point in the last century, some of these mosquitoes found their way into the London Underground system where they survived very well, feeding on the blood of commuters. Recent study of these 'underground mosquitoes' has found that they show very different behaviour patterns to their overland counterparts, and will not willingly interbreed with them. The 'underground mosquito' is now classified as a different species – *Culex molestans*.

An Inexact Science

We humans love to categorize things, and when it comes to the world of animals, the species is the preferred unit of organisation. We define them by whether or not they can breed with each other – horses and donkeys are clearly similar and will mate, but their offspring is infertile so we can safely consider them different species. Go back in time far enough, though, and you will find an animal that is an ancestor to both horses and donkeys, and there is no exact moment when the two lineages split. The yellow wagtails in Britain look very different to the ones in Greece, but they will still successfully interbreed so they are not separate species – yet. Give it a few more eons of separation and they could well be. At the moment they are classed as separate subspecies.

The division of a species into two species is called a

speciation event. Because huge timescales are involved, it's very rare for us to actually observe one, though we stand a better chance among animals which have lots of generations in a short space of time. Scientists have seen the appearance of new species of bacteria in the lab, including one with the handy ability to digest nylon. And by studying differences in DNA, we can determine that the most recent ancestor that we share with our closest living relative – the chimpanzee – lived about five to seven million years ago.

STASIS

Environmental change means species must adapt or die. However, sometimes there is little or no change in a certain environment for a huge amount of time, so the pressure on the organism is to stay just as it is. The result can be what's known as a 'living fossil' – an organism that looks pretty much the same as its ancient ancestors. The best-known example is a big, ungainly fish called the coelacanth. Modern coelacanths haven't changed much from their fossil relatives from before the time of the dinosaurs, more than 400 million years ago.

POPULATION

There are currently well over six billion members of the species *Homo sapiens* walking about on planet Earth. Haven't we done well? In sharp contrast, the total wild population of Spix's macaw, a pretty blue parrot from Brazil, was – until recently – one*. The vast majority of species that have ever existed are now extinct. While the process of evolution is the means by which species adapt to changes in their environments, it takes time, and when environments change very quickly most species don't adapt in time, especially when their population is small.

WHAT REGULATES POPULATIONS?

Populations of living things are constrained by the resources available in their environment. A 'resource' could mean something as simple as physical space (think of barnacles on a rock, or trees in a dense forest). It also covers food, water, suitable breeding sites, places to hide from predators and anything else the species in question needs to survive. The environment will have enough resources to sustain a certain population size of each species – this is its 'carrying capacity'. The population won't grow beyond the

---

*It's now none. There are a few dozen in captivity still.

environment's carrying capacity because some individuals will die or fail to breed due to lack of resources.

WHAT CHANGES POPULATIONS?

Environmental changes will affect the carrying capacity of that environment with regard to each species that lives in it. The knock-on events can be simple – a drought strikes, there's less water to go around so animals die of thirst, and the population stabilizes at a lower level. It may work the other way round too – as the agricultural revolution spread west across Europe, the newly created farmland habitats allowed the populations of birds like the collared dove, which favours open habitats, to increase and spread. Of course, in a complex and dynamic system like planet Earth, things can get complicated and consequences can be difficult to predict. For example, a heathland fire may kill some of the animals that live there, but may also create new and better habitat for the survivors so numbers bounce back quickly.

BOTTLENECKS

In species that reproduce sexually, reducing the population too much causes inbreeding depression, when all of the individuals able to breed are very closely related. This is bad news, because inbred organisms are more likely to have health problems – the lack of genetic variety means that a disease affecting one individual is likely to also affect the others. The population of the northern elephant seal fell to just thirty individuals in the 1890s – though numbers have since recovered, all of the thousands of northern elephant seals alive today are very closely related to each other.

Another species to have gone through a population bottleneck like this is *Homo sapiens*, i.e. we humans. Some 70,000 years ago a supervolcano in Indonesia erupted, causing a mini ice age and killing off a huge proportion of the humans on Earth at the time, leaving no more than 10,000 potential breeding couples (possibly far fewer). Today, the six billion humans on Earth exhibit a lot less genetic variability than the 150,000 or so chimps and 125,000 gorillas.

PREDATION

All animals consume other organic material. Many eat just plants – the herbivores – and some eat both plants and animals – the omnivores. Then there are quite a lot of them that only eat other animals, which they catch and kill. These are the carnivores or predators. When we think of a predator, we probably imagine something big, scary and well-equipped with pointy claws and teeth. However, a robin is as much a predator of worms and spiders as a leopard is a predator of antelopes.

THE FOOD WEB

You probably remember the concept of a food chain. Here's an example.

Grass > Rabbit > Fox

The rabbit eats the grass, the fox eats the rabbit. Let's flesh it out (so to speak) a bit more.

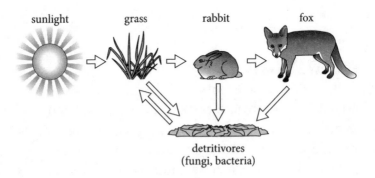

detritivores
(fungi, bacteria)

We now have a web rather than a chain, showing how energy and nutrients cycle from organism to organism. The grass gets its energy and nutrition from the sunlight and the earth. The rabbit eats the grass, the fox eats the rabbit. The fox has no predators, but when it dies its body is broken down by decaying organisms, or detritivores, like fungi and bacteria. The same thing happens to grass and rabbits that die rather than get eaten. The detritivores release nutrients in a form that the grass can take up. You can add as many other organisms as you like to a food web – this one, set in typical English countryside, could easily include such things as oak trees, stinging nettles, assorted insects, flycatchers, sparrowhawks, wood mice and badgers.

PREDATOR–PREY RELATIONS

It's fairly obvious that, with the simple food web shown opposite, if you take away the rabbits, you can't have any foxes. For that reason, the foxes can't wipe out the rabbits. Because each fox needs lots of rabbits, the foxes will die out before they have eaten every last rabbit. In real life, of course, there aren't just rabbits, there are mice and birds and worms and many other things that the foxes can eat instead, but it is still very unlikely that a predator will ever kill off every last individual of any of its prey species.

If rabbits became rare, the foxes that survived best would be the ones that were good at catching prey other than rabbits.

In the average ecosystem, predators and prey have evolved together over millenia, and their populations fluctuate in a predictable pattern. In the Arctic, snowy owls mostly eat lemmings. In years of high lemming population numbers, most of the baby owls will get enough to eat and the owl population will grow. If lemming numbers crash, a lot of baby owls will starve. Over time, their numbers fluctuate like this:

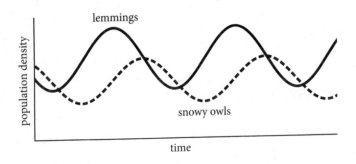

If we took away the owls, the lemmings' line of the graph may go up higher, if their numbers have not already reached the carrying capacity of the habitat. However, in natural situations, predators' numbers are much more strongly controlled by prey numbers than the other way round.

Predators that are moved from their native lands to new places can have much more damaging effects. In a bid to control the rampant population of introduced rabbits in Australia, people introduced stoats to eat them. A sensible enough move, you might think – stoats are expert rabbit hunters in Europe after all. However, the introduced stoats were soon happily decimating the small native animals, none of which had ever co-existed with stoats so, unlike the rabbits, had not evolved any kind of defence against them.

EXTINCTION

We have already seen that most of the species that ever lived are already extinct. Life has existed on Earth for about 3.7 billion years and over that time things have changed here a great deal. We have had ice ages, supervolcano eruptions, asteroid strikes and plenty of tectonic activity changing the shapes of our continents and causing earthquakes and other mayhem in the process. All of these events have placed massive pressure on Earth's living passengers, and have put paid to the existence of more than 99% of all species, ever.

Defining extinction

This is one of the simpler concepts to get your head around in biology – when every last member of a species has died, that species is extinct. Species that are not extinct are called extant. If there are some individuals of a species left but they are not able to breed (for example, they are all males) that species is functionally extinct. Several species are extinct in the wild but have living representatives in captivity – Spix's macaw, which we mentioned in the Population section, is an example.

Causes of Extinction

There is an urban legend that a small flightless bird called the Stephens Island wren was completely exterminated within a few weeks by a single cat brought to the island by sailors in 1894. While the veracity of some details of the story are disputed, there's no doubt that this particular environmental change was too fast for the small wren population to adapt before all were gone. Such is the cause of all extinctions – a change too fast and too severe for the species to adapt to, whether it be the arrival of a new predator, a climate change, a volcano eruption, the destruction of a large swathe of habitat to build a new hotel or any number of other disasters. If a large part of a population is destroyed, the remaining populations may be too fragmented to connect up again and too small to survive alone.

MASS EXTINCTIONS

The fossil record shows us that since life began on Earth there have been five periods of time when extinction rates went up way beyond the usual level. These mass extinctions include the one sixty-five million years ago, which killed all of the dinosaurs, and an even more devastating event 250 million years ago which wiped out 96% of all marine species. Some biologists say that the current high rate of extinction due to human activity warrants inclusion as the sixth mass extinction.

GENETICS

CHROMOSOMES

The DNA that lives in the nucleus of each cell in your body* is one long molecule mixed up with proteins and is called chromatin. When a cell is about to divide, the chromatin is organized into strands called chromosomes. When they are about to divide they look like tall thin Xs, with a join in the middle where they separate. Humans have forty-six of them, in twenty-three pairs. The pairs all match apart from one – in men only. In women both of the 'sex chromosomes' are X chromosomes, but in men one is X and the other is the much smaller Y. A full set of chromosomes is called a karyotype. You might have seen a karyotype displayed with the chromosomes in pairs. If you haven't, the human male karyotype looks like the illustration opposite.

Each chromosome is made of numerous genes –

---

*Apart from the ones that don't have nuclei, like your red blood cells.

a gene is a length of DNA, which holds the code for making a specific protein, as well as coding for whether it is 'switched on' or 'switched off'.

The chromosomes contain all the instructions you need to make all the proteins that form your adult body. From the moment of fertilization, the chromosomes in the single cell, which results from the fusion of egg and sperm, are at work coding new proteins that make new cells and carry the instructions for what the cells need to do. Because half your chromosomes come from your mother and half from your father, when put together they make a brand new combination of genes.

WHEN CHROMOSOMES DIVIDE

As we saw in the Reproduction section, there are two kinds of cell division. There's mitosis, when a cell simply duplicates itself and all its chromosomes, and meiosis, which happens with sperm and egg cells, producing from each parent cell four daughter cells with half the chromosomes of their parent cells. These cells are called haploid, and each contains twenty-three chromosomes, one of each pair.

During meiosis, the cells divide twice, so one cell becomes four. On the first division, the two new cells still have forty-six chromosomes in twenty-three pairs, but the chromosomes themselves are rearranged – a process called recombination. So bits of the first member of chromosome pair 1 get combined with bits of the second member of pair 1, and so on, creating two cells with a different set of genes to their parent cell. Then the second division occurs, resulting four haploid cells. The recombination means that no two haploid cells ever contain the same combination of genes.

INHERITANCE

Most of us can see some of our parents' features in our own faces. Most of our features are shaped by the influence of many genes (we have something like 20,000 genes altogether). To understand how inheritance works,

PHENOTYPE AND GENOTYPE

Time to explain these two tricky terms. Your phenotype is your outward appearance. In the case of freckles, your phenotype would be 'freckly' or 'not-freckly'. Your genotype describes which alleles (see p 182) you have with regard to that particular trait – whether you have two f alleles, two F alleles or one of each. If both your copies of an allele for a trait are the same, your genotype is homozygous for that trait – if they are different, you have a heterozygous genotype for that trait*. As you'll see, it's not always possible to tell a genotype by looking at the phenotype.

*Homo = the same and hetero = not the same.

it's easiest to look at a trait controlled by a single gene – the trait is either present or absent depending which versions – or alleles – of a particular gene you have. Traits like this are called Mendelian, after their discoverer Gregor Mendel. Common Mendelian traits in humans include cleft chin, facial freckles (most obvious in pale-skinned people, but anyone can have them), the ability to flex your thumbs backwards and whether you have free or attached earlobes*.

ALLELES

A Mendelian trait is either dominant or recessive. If it's dominant, the trait will be expressed whether you have one or two copies of the allele. If it's recessive, you need two copies. Let's take the freckles example. This gene resides on chromosome 4, so on each of your copies of chromosome 4 you have either the dominant having-freckles allele or the recessive no-freckles allele. Let's call them F and f respectively.

Bob and Mary both have freckles, Jim and Susan both lack them. From that we can deduce that both Jim and Susan have the genotype ff – two copies of the no-freckles allele.

$$Jim = ff$$
$$Susan = ff$$

---

*i.e. 'dangly' earlobes or earlobes joined straight to the side of the head, if you want the non-technical terms.

With Bob and Mary, we can't be so confident. They have the same phenotype – freckly, but because F is dominant over f, they could have different genotypes – either one copy of F and one of f, or two of F.

$$Bob = FF \text{ or } Ff$$
$$Mary = FF \text{ or } Ff$$

If Jim and Susan have babies, they will all have the same non-freckly phenotype and the same ff genotype as their parents – because neither Jim nor Susan have any F alleles to contribute. If Bob and Mary both happen to be FF, they too will have children with the same phenotype (freckly) and genotype (FF) as themselves.

$$ff \text{ parent} \times ff \text{ parent} = 100\% \text{ } ff \text{ offspring}$$

$$FF \text{ parent} \times FF \text{ parent} = 100\% \text{ } FF \text{ offspring}$$

What if Bob is FF and Mary is Ff, though? Because each of their children will inherit an F allele from Bob, they will all have the freckly phenotype. But because half of them will get Mary's F allele and the other half her f allele, they will be half FF and half Ff.

$$FF \text{ parent} \times Ff \text{ parent} = 50\% \text{ } FF, 50\% \text{ } Ff \text{ offspring}$$

The most interesting outcome is if both Bob and Mary are *Ff*. There are four possible combinations from this – *FF*, *Ff*, *Ff* the other way around, and *ff*.

> *Ff parent x Ff parent*
> *= 25% FF, 50% Ff and 25% ff offspring*

To explore the final possible combinations we need our couples to do a bit of clandestine wife-swapping. Let's say we've established that Bob is *FF* and Mary is *Ff*. Bob and Susan's love-children will each inherit one of Bob's two *F* alleles and one of Susan's two *f* alleles, so they will all have a freckly phenotype and an *Ff* genotype.

> *FF parent x ff parent = 100% Ff offspring*

Half of Jim and Mary's children will inherit Mary's *F* allele and the other half will inherit her *f* allele, while all will get one of Jim's two *f* alleles.

> *Ff parent x ff parent = 50% Ff and 50% ff offspring*

From this we can deduce that (assuming both illicit pairings result in a child and everyone concerned knows about Mendelian inheritance) Bob and Susan's affair has a 100% chance of being detected, while Jim and Mary's indiscretions have only a 50% chance of discovery.

INHERITED DISEASES

Not all Mendelian traits are innocuous things like freckles. Some devastating diseases are passed on this way too. One is Huntington's chorea, a slowly progressive and fatal disease of the nervous system which develops in early middle-age. It is a dominant trait, meaning that anyone who has a copy of the allele will develop it and anyone with one affected parent has a 50% chance of developing it. Genetic testing means that problems like this can be detected and affected people can make an informed decision whether to have a family. Cystic fibrosis is a recessive trait, i.e. you can inherit it from two unaffected parents if they are both heterozygous for it.

Alleles on the X chromosome work in a slightly different way. The X and Y chromosomes are not a matching pair – the Y is much smaller with far fewer genes on it. So in men (who have one X and one Y) an allele for a recessive trait on the X chromosome will always be expressed in the phenotype, while in women (with two Xs) it may not. An example is the blood-clotting disorder haemophilia. If a man inherits an allele for this from either parent he will develop haemophilia, but a woman will only develop it if she inherits the allele from both parents*.

---

*Genetic disorders are not the same as congenital conditions. The latter are not inherited, but are caused by something going wrong with the chromosomes prior to or just after conception, e.g. Down's syndrome.

REPRODUCTION AND CLONING

As we've seen, in sexual reproduction genetic variety is introduced, both during meiosis when each parent's chromosomes get chopped up and mixed up (recombination) and at fertilization when half of the chromosomes from each parent are combined to form a brand new full set.

The alternative to sexual reproduction is cloning, where a single parent produces an exact copy of itself, with the same chromosomes. What little variety there is between parent and offspring can only be introduced in the form of genetic mutations that may or may not occur.

The cells in our body reproduce by cloning, and single-cell organisms do it too. Some more complex life-forms also retain the ability to naturally clone themselves, while retaining the option of sexual reproduction as well. In animals, this process is called parthenogenesis, and is easily observed among various kinds of insects. Problems with greenfly or blackfly in your garden? Those little aphids carpeting the rose stems are all clones, born alive and churned out by the females all through the summer with no need for any male involvement at all.

Cloning in Mammals

Scientists have been working on cloning mammals of various

species for decades. The first viable mammal clone was Dolly the sheep, born in 1996. Her creation began with the removal of an unfertilized egg from the ovary a female sheep – Sheep A. The nucleus was removed, and replaced with that from a cell from the mammary gland of a different female sheep – Sheep B. This new cell was placed in the uterus of yet another female sheep – Sheep C, I suppose, where it grew and developed. In due course Dolly was born, a clone of Sheep B. Since Dolly, scientists have successfully cloned various other mammals, including cats, cows, horses and rhesus monkeys, though the failure rate remains very high.

WHY CLONE?

Cloning has a strong whiff of 'Frankenstein science' about it, second only to genetic engineering (directly manipulating an organism's genetic makeup) in its ability to inspire unease. It's hard to imagine a non science-fiction scenario where it would find widespread acceptance. However, cloning does have its uses – real and potential. Cloned plants have been used in agriculture for many years. We could clone new organs and solve the problems of insufficient registered donors and incompatible hosts. Endangered species could be saved from extinction. We could even make a theme park full of dinosaurs … what could possibly go wrong … ?

Entries for illustrations are denoted in italics

abiogenesis 158, 163
acceleration 34, 43, 44
acid rain 85
adaptation 169
ADP 152–3
adrenaline 129
agricultural revolution 170
air 27, 83–8, 137, 139
air pollution 84–8, 112
alkali metals 61–2
alkalis 77
alleles 181–5
alloys 97–8
alveoli *138*, 138–9
amino acids 108, 131, 148, 158
amniotic sac 135, 136
amperes 27, 28
amplitude 47
anions 65
aorta 116
Archaea 160–2
argon 62, 84
arteries 115–17
asteroids 37
atomic mass 64, 66
atomic numbers 58, 61, 64, 66
atomic weight 66
atoms 63–74; Big Bang 41; chemical reactions 74; crystals 72–3; electricity 32; electron shells 62; elements 58; half-life 55; physics and chemistry 14; radioactivity 53, 54
ATP 152–3
atria 116, *117*, 118
Avogadro's hypothesis 44
axons 127, *127*

babies 135–6
bacteria 159–62; colon 132; enzymes and 149; new species 168; principal life forms 146; skin function 124; white blood cells and 115
bases 77
Big Bang 41–2
birds 19, 23, 111, 132
black holes 39
bladder 132
blind spots 142, *143*
blood 114–18, 122- see also red blood cells; white blood cells

blue-green algae 150–1, 161
bonds: chemical 67–73, 153; covalent 69–70, 71, 83; double 105; electron 80; hydrogen 71, 73, 76, 103; ionic 69–70, 75; metallic 71, 96; molecules 71, 80; muscles 123
bones 119–22, 123
bowels 110
Boyle's Law 43, 44
brain 128, 144
breathing 137
bridges 99, 101

caesarean section 135
cancer 49, 52, 53, 54
capillaries 115, 116, 131, 139
carbohydrates 103, 104, 107, 109, 131, 151
carbon: atom *64*, 67, 69, 103; crude oil 89; fatty acids 105–6; isotopes 66; moles 79; radioactive form 55; rust 75–6; steel 97–8, 100
carbon dioxide: air pollution 85; alveoli 139; atmospheric 84, 139; fossil fuels 19; galvanized steel 100; methane 74; molecule 68–9, *69*; photosynthesis 150, 151; plants and animals 153; rust 76
carbon monoxide 85, 112
cardiovascular system 114
carrying capacity 169–70
cartilage 121
catalysts 81–2, 159
cations 65
cell biology 146–56
cells 146–50; biology 15; blood 115, 118, 139, 146, 178; bones 122; brain 128; cell division 135, 136, 164, 180; DNA 164, 178; Dolly the sheep 187; Down's syndrome 186; ears 144; endosymbiosis 159–60; eukaryotic 160–2; eyes 142; fertilization 179; haploid 180; hormones 154, 155, 156; nerves 126–7; oxygen 114; plants *149*; reproduction 164; ribosomes 162; senses 140, 145; types 161
cellulose 104, *104*, 109–10
central nervous system 128
cerebellum 128
cerebral cortex 128
cervix 135
Charles's Law 43–4
chemical reactions 67, 73–82
chimpanzees 168
chlorine 60–1, 70, *70*, 109
chlorofluorocarbons (CFCs) 86

chlorophyll 151
chloroplasts 149, 150, 151, 160–2
chromatin 178
chromium 98, 100
chromosomes: 178–80; cell division 135; diagram *179*; fertilization 136, 186; freckles gene 182; X and Y 185
circulation 114–19
clitoris 134
cloning 186–8
coelacanths 168
collagen 122
collision theory 78, 80
colon 109, 130, 132
comets 37
compounds 60, 73, 76–7, 81, 98, 152
conduction 23, 25, 26–7, *26*, 96, 97
construction materials 98–101
convection 23, 25, 26, *26*
core (Earth's) 93, 94
covalent bonds 69–70, 71, 83
cows 132–3
crops (stomachs of birds) 132
crude oil 88–90
crust (Earth's) 93, 94–5
Culex mosquitoes 166–7
Curie, Marie 54–5
Curie, Pierre 54–5
current 28, 30, 48
cyanobacteria 150–1, 161
cystic fibrosis 185

DDT 111
diabetes 156
diesel 90
diets 107–10
diffusion 79, 87
digestive system 122, 130–3
distillation 89
DNA: adenosine 152; Archaea 160, 162; cell nucleus 148; chloroplasts and mitochondria 161; mutation 164; nature of 178; polymers 103–4; RNA 158–9; species 168
Dolly the sheep 187
Doppler effect 40
Down's syndrome 186
duodenum 132
dwarf stars 38–9

e=mc² 20–1
ears 144, 145
Earth 93–5

eggs 134, 136, 179, 187
Einstein's Theory of Special Relativity 20
ejaculation 136
electricity 18–22, 27–31, 32, 96, 97
electromagnetic force 32–3, 48
electromagnetic spectrum 25, 48–53
electronegativity 72
electrons: bonds 80; carbon *64*, 67; chlorine 70, *70*; electricity 32; fluorine 72; helium 68; hydrogen 42, *68*; ions 65; nitrogen 83; noble gases 62; orbiting their atoms 63; 'seas of' 71; shells 62, 67, 68, 69, 70, 73; sodium *70*; valency 69; voltage 28
elements 58–62
embryos 136, 155
endocrine glands 155
endosymbiosis 148, 150, 159–62
energy 18–22; activation 80; cell's 148; chemical reactions 76–7; glucose 105; photosynthesis 150; respiration 152–3; waves 45–6
entropy 24
enzymes 131, 149, 159
epidermis 124–5
equations 30, 34, 44, 74, 151, 152
eukaryotic cells 151, 160–2
evolution 160–4, 169
extinction 169, 175–7
eyes 140–4, *141*, 143

Fallopian tubes 134, 136
fallout shelters 56
fats 105–6, 107, 125, 130–1, 154–5, 158
fatty acids 105–6, *106*, 108
fertilization 134–6, 179, 186
fibre (dietary) 109–10, 132
fission 19, 20
fluorine 62, 66, 72
foetuses 136
food webs 172–3
forces 14, 32–5, 43
fossil fuels 18–19, 21–2, 85, 88, 89
freckles 181–5
frequency 47
fur 125

g 34
galaxies 35–6, 39, 40, 41
gall bladder 130, 131
galvanized steel 100
gamma radiation 49, 52–3, 54, 56
GCSEs 14
generators 22, 28

genes 156, 178–9, 180–2, 185
genetic engineering 188
genetic mutation 163
genetics 178–88
genome 164
genotypes 181, 182–3, 184
geothermal power 21, 22
gizzards 132
glucose 105, 151, 152, 155–6
gold 60, 95, 97
gravitational force 33–4
gravity 34–5, 38, 39, 93
greenhouse gases 85

haemoglobin 112, 115
haemophilia 185
half-life 55
halogens 62
haploid cells 180
heart 116–18, *117*, 137
heat 22–7, 96
helium 38, 42, 60, 64, 68, 84
hertzes 47
hormones 154–6
houses 99
Huntington's chorea 185
hydrocarbons 76, 85, 89–90, 103
hydroelectric power 21–2
hydrogen: alkalis 77; atomic number 64; bonds
    71, 73, 76, 103; chemical symbol 60; crude
    oil 89; Earth's atmosphere pre-life 158; fatty
    acids 105–6; fluoride 72, 80; formation 42;
    polymers 103; simplest element 68; stars 38;
    valency 69; wood 76
hypotheses 16

Iceland 22
ileum 132
infrared radiation 25, 51
inheritance 180–6
insulation 26–7
insulin 155
interstitial fluid 118
ionic bonds 69–70
ions 65, 74, 86, 147
iron 75–6, 93–4, 95, 97–8, 109
ISO 14000 91
isotopes 66–7, 79

jejunum 132
joints 121

karyotypes 178
kidneys 130, 132, 155
kilowatt hours (kWh) 30
kinetic energy 19, 21, 22- *see also* energy

laptops 31
laws of physics 42–4
lead 56, 90
life cycle assessments (LCAs) 90–2
light 48, 50, 51, 140, 142, 150
liver 108, 110, 130, 131, 137, 155
lungs 137–8, *138*
lymphatic system 118

magnetism 33
mantle 93, 94
mass 34–5, 43
mathematics 14
matter 20
meiosis 135, 180, 186
melanin 124, 125
Mendel, Gregor 181–2, 184, 185
Mendeleyev, Dimitri 61
menstruation 134, 154
metabolism 154
metallic bonds 71, 96
metalloids 96
metals 23, 69, 93–8, 100, 101
meteoroids 37
methane 74, 85, 158
microwaves 51
Milky Way 35
minerals (dietary) 109
mitochondria 148, 160–2
mitosis 135, 180
molecules: Avogadro's hypothesis 44; bonds
    between 71, 80; carbon dioxide 68–9,
    69; chemical reactions 80; chemistry 14;
    electromagnetic force 32; ice and water 76;
    ions 65; polar 72; polymers 102
moles (units of quantity) 79
monomers 102, 104
Moon, the 35, 40
moons 37, 39
mother's milk 135
muscles 122–3, 155
mutation 163–8, 186

National Grid 29
natural selection 163–8
nervous system 126–9, 140, 185
neurons 126, 128

neurotransmitter 127
neutrons: Big Bang 42; carbon atom *64*, 66, 67; nuclear power 19; nucleus of atom 63; strong and weak nuclear forces 33
Newton, Isaac 43
nickel 94
nitrogen 83, 137, 160
nitrogen dioxide 85, 87, 112
noble gases 62, 68, 84
noble metals 95, 97
nuclear decay 52, 53–5
nuclear power 19, 21–2, 56, 67
nucleotides 103, 158
nucleus (atom) 53, 54, 63, 67
nucleus (cell) 127, 135, 148, 156, 162, 187
nutrition 107–10

oesophagus 130, 132
oil 98–100
organelles 146, 148, 149–50, 160, 162
organic chemistry 102–10
osmosis 79
ovaries 134, 155, 187
oxidation 61, 75–6
oxygen: air 83, 139; alkali metals 61; alveoli 139; blood circulation 114–17; breathing 137; carbohydrates 103; carbon dioxide 69; Earth's crust 95; energy release 152; methane 74; photosynthesis 150, 151; plants and animals 153; rust 75; sodium hydroxide 77; valency 69
ozone 86

pancreas 130, 131, 155
parthenogenesis 187
particles 14, 48, 54
penis 134
peppered moths 165
peregrine falcons 166
periodic table 58–62, *59*, 94, 96
pesticides 111
petrol 89–90
pH scale 77
phenotypes 181, 183, 184
photons 49, 50, 151
photosynthesis 150–3
placenta 136
Planck's Constant 49
planets 35, 36, *36*, 39, 42
plant cells 149–50
plasma (anatomical) 115
plasma (stellar gas) 38, 65
platelets 115

poisons 110–12
polymers 102–6, *102*
polysaccharides 105
population 169–71, 175, 176–7
power stations 21, 28–9
predation 171–5
prostate 134
proteins 103, 107–9, 123, 131, 147–8, 179
protons: atomic mass 66; atomic number 58; atoms 32; carbon atom *64*, 66, 67; helium and hydrogen 42; ions 65; nuclear forces 33; as nucleus of atom 63
pseudoscience 16

quantum mechanics 49

radiation 25, *26*, 26, 65- *see also* radioactivity
radio 49, 50
radioactivity 53–6, 66, 67
rainbows 51
reaction rates 80–2
red blood cells 115, 139, 146, 178
redshift 40
refraction 51
reproductive system 133–6
resistance 30
resources 169–70
respiratory system 137–9
ribosomal RNA 148, 158–9
ribosomes 148, 162
rumination 132–3
rust 75–6, 97

salts 77
sampling 87–8
scrotum 133
sebaceous glands 125
semen 134
sensory receptors 126
sensory systems 140–5
sexual intercourse 133, 136
silicon 95, 96
skeletal structure 119–22
skin 122, 124–6, 144
small intestine 130, 131–2
smell 145
smog 85
sodium 60–1, 70, *70*, 77, 95, 109
sodium chloride 61, 77, 78
sodium hydroxide 77
solutions 78–9
sound waves 47

space 34, 40, 41
species: endangered 188; extinct species 169, 175, 176; population 171; predators and prey 173; preferred unit of organisation 167–8
spectrum 25, 48–53
sperm 134, 136, 179
spinal cord 120, 128
stainless steel 98, 100
starches 105
stars 35, 38–42
steam 22
steel 97–8, 100
stomach 130, 131
stone 99–100
strong nuclear force 33
subatomic particles 14
sugars 103, 104–5, 131, 151, 155, 158
sulphur dioxide 85, 112
Sun, the 35, 38, 40, 150
supernovas 39
sweating 125, 126

taste 145
tendons 123
testes 133–4, 155
testosterone 155
thermals 23, 25
thermodynamics 24–5
tides 21
tongue 131
trans fats 106
transformers 29
triethanolamine 87
turbines 19, 22

ultrasound waves 47–8
ultraviolet light 52, 124
universe, the 35–42, 157
urethra 132
uterus 134, 135, 136, 187

vacuoles 150
vagina 134
valency 69
valves (heart) 116, *117*, 118
vas deferens 134
veins 116, 117
velocity 44
venom 112
ventricles 116, *117*, 118
vertebrae 120, 128
vibration 46
viruses 159
vitamin D 124
vitamins 109
volcanoes 21, 85, 94, 158, 171
voltage 27–30

water 72–3; chemical compound 69; food 51; hydrogen bonds 71; methane 74; pH value 77; photosynthesis 150; as solvent 78–9; sustainable power source 21; testing drinking water 88; waves 45
water vapour 84, 158
watts 28
wavelength 47, 50
waves 25, 45–8, *46*
weak nuclear force 33
weight 34–5
white blood cells 115, 118
wind power 19
womb- *see* uterus

X chromosomes 178, 185- *see also* chromosomes
X-rays 52

Y chromosomes 178, 185- *see also* chromosomes

zinc 100, 109

FURTHER READING

A Brief History of Time by Stephen Hawking (Bantam Books)

The Blind Watchmaker by Richard Dawkins (Penguin)

Bad Science by Ben Goldacre (Harper Perennial)

Cosmos by Carl Sagan (Abacus)

The Chemistry of Life by Steven Rose (Penguin)

On the Origin of Species by Charles Darwin (Oxford University Press)